SpringerBriefs in Electrical and Computer Engineering

Series editors
Woon-Seng Gan, School of Electrical and Electronic Engineering, Nanyang
Technological University, Singapore, Singapore
C.-C. Jay Kuo, University of Southern California, Los Angeles, CA, USA
Thomas Fang Zheng, Research Institute of Information Technology, Tsinghua
University, Beijing, China
Mauro Barni, Department of Information Engineering and Mathematics, University
of Siena, Siena, Italy

SpringerBriefs present concise summaries of cutting-edge research and practical applications across a wide spectrum of fields. Featuring compact volumes of 50 to 125 pages, the series covers a range of content from professional to academic. Typical topics might include: timely report of state-of-the art analytical techniques, a bridge between new research results, as published in journal articles, and a contextual literature review, a snapshot of a hot or emerging topic, an in-depth case study or clinical example and a presentation of core concepts that students must understand in order to make independent contributions

More information about this series at http://www.springer.com/series/10059

Miao Pan • Jingyi Wang
Sai Mounika Errapotu • Xinyue Zhang
Jiahao Ding • Zhu Han

Big Data Privacy Preservation for Cyber-Physical Systems

Springer

Miao Pan
Department of Electrical and Computer
Engineering
University of Houston
Houston, TX, USA

Sai Mounika Errapotu
Department of Electrical and Computer
Engineering
University of Houston
Houston, TX, USA

Jiahao Ding
Department of Electrical and Computer
Engineering
University of Houston
Houston, TX, USA

Jingyi Wang
Department of Electrical and Computer
Engineering
University of Houston
Houston, TX, USA

Xinyue Zhang
Department of Electrical and Computer
Engineering
University of Houston
Houston, TX, USA

Zhu Han
Department of Electrical and Computer
Engineering
University of Houston
Houston, TX, USA

ISSN 2191-8112   ISSN 2191-8120   (electronic)
SpringerBriefs in Electrical and Computer Engineering
ISBN 978-3-030-13369-6   ISBN 978-3-030-13370-2   (eBook)
https://doi.org/10.1007/978-3-030-13370-2

Library of Congress Control Number: 2019933298

This Springer imprint is published by the registered company Springer Nature Switzerland AG.
The registered company address is: Gewerbestrasse 11, 6330 Cham, Switzerland

Preface

Cyber-physical systems (CPS) often referred as "next generation of engineered systems" are sensing and communication systems that offer tight integration of computation and networking capabilities to monitor and control entities in the physical world. The advent of cloud computing technologies, artificial intelligence, and machine learning models has extensively contributed to these multidimensional and complex systems by facilitating a systematic transformation of massive data into information. Though CPS have infiltrated into many areas due to their advantages, big data analytics and privacy are major considerations for building efficient and high-confidence CPS. Many domains of CPS, such as smart metering, intelligent transportation, health care, sensor/data aggregation, crowdsensing, etc., typically collect huge amounts of data for decision-making, where the data may include individual or sensitive information. Since vast amount of information is analyzed, released, and calculated by the system to make smart decisions, big data plays a key role as an advanced analysis technique providing more efficient and complete solutions for CPS. However, data privacy breaches during any stage of these large-scale systems, either during collection or big data analysis, can be an undesirable loss of privacy for the participants and for the entire system.

This book focuses on effective big data analytics for CPS and addresses the privacy issues that arise in various CPS applications. Because of their numerous advantages, CPS and its communication networks inevitably become the targets of attackers and malicious users either during data collection, data storage, data transmission, or data processing and computation, keeping users' information at risk. Given these challenges, this book endeavors to develop a series of privacy-preserving data analytic and processing methodologies through data-driven optimization based on applied cryptographic techniques and differential privacy and focuses on effectively integrating the data analysis and data privacy preservation techniques to provide the most desirable solutions for the state-of-the-art CPS with various application-specific requirements.

Houston, TX, USA   Miao Pan
Houston, TX, USA   Jingyi Wang
Houston, TX, USA   Sai Mounika Errapotu
Houston, TX, USA   Xinyue Zhang
Houston, TX, USA   Jiahao Ding
Houston, TX, USA   Zhu Han
December 2018

Contents

Acronyms

3DPP	Data-Driven-Based Spectrum Trading Scheme with Secondary Users' Differential Privacy Preservation
AP	Access Points
CP	Content Provider
CPS	Cyber-Physical Systems
CR	Cognitive Radio
CRN	Cognitive Radio Network
DCA	Descending Clock Auction
DDP	Distributed Differential Privacy
DP	Differential Privacy
EDR	Emergency Demand Response
FCC	Federal Communications Commission
ICN	Information-Centric Networking
LDP	Local Differential Privacy
MVNO	Mobile Virtual Network Operator
OLH	Optimized Local Hashing
PCS	Paillier Cryptosystem
PPCA	Privacy-Preserving Clock Auction
PSP	Primary Service Provider
PU	Primary User
QoS	Quality Of Service
RA-SP	Risk-Averse Stochastic Optimization Approach
SG	Smart Grid
SSP	Secondary Service Provider
STED	Secondary Traffic Estimator and Database
SU	Secondary User

Cyber-Physical Systems

1

Abstract

In this chapter, we first introduce the concept of the cyber physical systems (CPS) and explain the importance of privacy and security in CPS. Then, we present four different applications of CPS in the real world and briefly summarize the state-of-art research and the research challenges for the privacy preservation and big data analysis in these CPS systems. In this chapter, we provide content outlines for the rest of the book.

1.1 Introduction to Cyber-Physical Systems

CPS are a sensing and communication systems that offers tight integration of networking and computation components to monitor and control entities in the physical world. CPS are largely referred to as the next generation of engineered systems that achieve the goals of stability, performance, robustness, and efficiency through effective integration of communication, computation, and control. CPS primarily consists of two entities: a physical system and a cyber system. The physical system constitutes of the embedded computers and sensors that collect measurements to describe the current state of the physical system, and then send them to the cyber system through communication networks in real time. Simultaneously, the cyber system processes the received measurements and obtains the status of the physical system. Depending on the processed results, the cyber system directs the physical system, achieving better performance and maintaining system stability.

© The Author(s), under exclusive license to Springer Nature Switzerland AG 2019
M. Pan et al., *Big Data Privacy Preservation for Cyber-Physical Systems*,
SpringerBriefs in Electrical and Computer Engineering,
https://doi.org/10.1007/978-3-030-13370-2_1

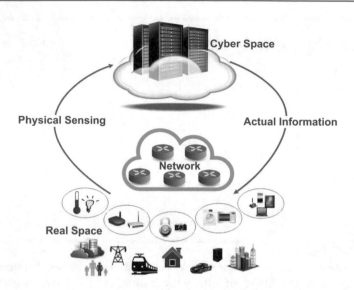

Fig. 1.1 Cyber-physical systems

   CPS have infiltrated into many areas, such as aerospace, energy, health care, manufacturing, transportation, and consumer appliances (Fig. 1.1). One of the major challenges for the CPS is big data analysis. Since CPS are often distributed across wide areas and typically collect huge amounts of information for data analysis and decision making, data privacy is a serious concern. For example, the operation of large-scale monitoring and control systems, such as smart grids relies on information continuously provided by their users. Collecting such information helps the system make smart decisions through big data analysis. By effectively analyzing the data through big data analysis the CPS can make the system more stable, and have better efficiency, scalability and resiliency. However, dealing with the uncertainty of the data is extremely challenging. For instance, when system collects the data from sampled users, the result cannot be 100% accurate to represent the whole user's characteristics. There will always be a gap between the sampled result and real result. Therefore, this work studies on mathematical relation between the sampled result and real result based on data driven optimization to effectively analyze data in various CPS. Apart from big data analysis, privacy is a serious concern in CPS and breaches in these applications can be an undesirable loss of privacy for the participants in these heterogeneous systems, thereby putting their promised benefits at risk. Data breaches in such large-scale systems can occur at any stage of the system either during data collection, data transmission, data operation, or data storage. Due to their numerous benefits and vast importance, CPS and its communication networks inevitably become the targets of attackers and malicious users. With rapid technological development, the need for new theories and tools that can effectively protect the individuals while collecting dynamic data in these large-scale distributed systems is increasing.

Therefore, in this book, we jointly consider these two problems and address these concerns for different CPS scenarios. This book focuses on the big data analysis and privacy of CPS, including its communication networks. Specifically, we present the big data analytics and privacy works in four applications of CPS. In the following sections of this chapter, we will briefly state each problem, and describe the contribution of each work.

1.2 Cognitive Radio Networks in CPS

Cognitive radio networks play an important role in CPS as a typical communication networks. Cognitive radio networks have created an efficient communication paradigm to enhance the utilization of the wireless spectrum. Within the traditional spectrum regulation framework, spectrum bands are exclusively assigned to wireless services by the Federal Communications Commission (FCC), resulting in inefficiency of spectrum utilization spatially and temporally. Majority of the wireless spectrum is licensed but under-utilized, and the unlicensed spectrum is over-crowded due to the rapid increase of wireless services. So, spectrum scarcity is becoming a major concern in recent days due to technical advances in the wireless networks and increased usage of smart devices.

Cognitive radio networks (CRNs) solve the above dilemma intelligently by allowing unlicensed users to opportunistically access the licensed bands providing that they don't interfere with licensed users. Typically, a cognitive radio network consists of two types of users: primary users (PUs) and secondary users (SUs). The secondary users are capable of sensing and adapting to their spectral environment. Specifically, by carefully sensing the primary user's presence and adjusting their own transmission to times when the primary user is idle, the secondary users can utilize licensed spectrum for transmission and therefore can dramatically improve spectral efficiency. When the primary user is present, the secondary user can either shut off their transmission or adjust their transmission parameters to guarantee an acceptable performance for the primary user.

The spectrum trading in CRNs is an efficient way of dynamic spectrum sharing to improve the inefficient spectrum utilization and meet the increasing demand for the spectrum. In spectrum trading, the PUS receive financial rewards by leasing their licensed spectrum to the secondary users that bid for the spectrum. Spectrum trading with SUs for monetary gains improves the inefficient spectrum utilization of PUs in the CRNs.

Despite those benefits, there are many challenges for pushing spectrum trading in practice. For example, due to hardware limitation of either PUs' or SUs' devices, they may have too limited sensing capability to know some spectrum trading opportunities nearby [8]; aiming to maximize the revenue, the PU may feel challenging to develop optimal selling strategies due to the SUs' traffic demand uncertainty; the SU may feel difficult to preserve its spectrum trading privacy (i.e., the SU's locations, true evaluation values of certain spectrum, traffic portfolio, etc.) [5, 15, 16], and so on. Those concerns may make PUs or SUs reluctant to participate in spectrum trading.

To facilitate PUs' and SUs' participation and make spectrum trading practical, recent studies [8] have introduced spectrum trading architectures based on existing wireless network infrastructure. Under those architectures, primary service provider (PSP) aggregates vacant spectrum bands from PUs, and sells the spectrum bands to secondary service provider (SSP) at wholesale price. The SSP will evaluate the spectrum supply uncertainty, make the spectrum purchasing decision, and further sell the purchased spectrum to SUs at retailed price. Here, the role of PSP/SSP can be played by base station in cellular networks, eNodeB in LTE networks, or mobile virtual network operator (MVNO), where the PSP/SSP has more sensing power than the individual SU. Although the spectrum trading architectures in [8] help to capture spectrum accessing opportunities, and the algorithms mathematically characterize spectrum supply uncertainty, it ignores the SUs' traffic demand uncertainty, which may have negative impact on PSP's revenue maximization. That is, without the accurate knowledge of SUs' traffic demands, the PSP cannot choose the optimal selling strategies to maximize its revenue. Moreover, the approach of using random variables to model the traffic uncertainty in previous research may be good enough to reflect the PUs' traffic patterns over a relatively long-term period, but it will not be able to represent SUs' traffic demands in real-time manner.

Therefore, following the framework of spectrum trading architectures, in this book, we further introduce a new entity, called secondary traffic estimator and database (STED), which is responsible for estimating the SUs' traffic demands in real-time manner and answering PSP's queries about SUs' traffic demands as shown in Fig. 1.2. Considering the large population of SUs in the PSP's coverage boundary, it is not efficient to crowdsource SUs' traffic demands by collecting each SU's demands in terms of time consumption and communication overhead. Thus, we propose to let the STED employ data-driven approach to collect sampled SUs' demands, construct reference demand distribution from sampled demands, and leverage reference distribution to estimate the demand distribution of all SUs.

Now, the leftover challenge hindering spectrum trading is the traffic privacy preservation of the sampled SUs. Taking the query procedure of SUs' demands shown in Fig. 1.2 as an example, the SU's traffic portfolio privacy is breached as

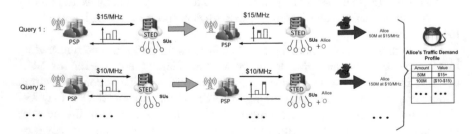

Fig. 1.2 Illustrative examples for the traffic demand privacy breach of SUs in spectrum trading

follows. For Query 1, the PSP will send a query about SUs' demand to STED, and the query is what the SUs' demand distribution is, if the price for spectrum accessing is $15/MHz. The STED will respond to this query with a traffic demand distribution of SUs at the cost of $15/MHz (e.g., 30% SUs would like to purchase 50M and 70% SUs would like to purchase 150M from 100 SUs in total). If a new SU, Alice, joins the group and she would like to purchase 50M at $15/MHz, the STED will update the SUs' demand distribution to the PSP's query (i.e., 30.7% SUs would like to purchase 50M, 60.3% SUs would like to purchase 150M from 100 SUs in total). From the differences of distributions, the PSP will derive that Alice would like to purchase 50M at $15/MHz or above. Through multiple queries, the PSP can easily learn Alice's traffic demand profile, which not only discloses Alice's true evaluation values of spectrum resources, but also classifies her personal traffic demands (e.g., voice, video, web browsing, social networking, online gaming, etc.) at different price levels.

In order to protect SUs' traffic demand differential privacy (DP) [9, 16, 28], in this book, we assume the STED is trustworthy, and entitle the STED to transform the SUs' demand distribution by adding noises before it responds to the PSP's queries. Instead of brutally hammering data-driven approach and DP together, we melt SUs' traffic demand DP into data-driven based spectrum trading, and mathematically prove its effectiveness. Based on that, we propose a novel data-driven based spectrum trading scheme with secondary users' differential privacy preservation, whose objective is maximizing the PSP's revenue.

1.3 Smart Grids in CPS

The evolution of existing electrical grids to smart grid (SG) is enhancing efficient power management, better reliability, reduced production costs, and more environmentally friendly energy generation. An indispensable part of this evolution, residential smart metering, brings smart grid into our homes, transforming them into smart homes of the future and allowing for more effective household energy monitoring and control. Smart grid's success heavily relies upon communication.

Although modernizing the electric grid introduces improvements, smart grid is also facing some challenges, where the most significant one is power outage/interruptions due to the supply and demand mismatch. As we know, a substantial amount of our electricity is generated from the coal, due to its affordable cost and huge coal reserves. In fact, coal-fired power generation takes relatively long time and the gas-fired power has flexible and prompt response to the real-time events in power grids [19]. To prevent power outage and satisfy customers' demand, the utility providers offset fluctuations by using more expensive gas-fired power or pumped-storage electrical power. For example, Siemens operates a number of gas-fired power plants all over the world, which is capable to provide flexible, reliable and efficient power supply.

Moreover, despite the great benefits from two-way flows of electricity and information in smart grid, the chances of malicious attacks and risks of privacy leakage increase. Smart metering is a promising solution to forecast and monitor electricity consumption of consumers. The smart meters are installed in each consumer's end (household, company, factory, etc.). The amount of electricity a customer used is measured and saved in an energy profiles, which will be sent to the utility provider at a requested time interval (the frequency can be as few as 1–5 min). The provider utility can predict the user's demand accurately, optimize the operation of all distribution resources, and improve the efficiency of the energy network. However, the energy profiles will be a potential target for well-motivated adversaries to compromise the customer's privacy. In this nearly real-time delivery of energy consumption profiles, the attackers can exactly observe the consumer's behavior, by comparing the differences between consumption profiles. For instance, the attackers/eavesdroppers can easily determine whether a consumer is at home by detailed energy consumption data, like TV or washing machine, and further surmise the consumer's house occupancy, meal times, working hours or lifestyle patterns.

There are some research efforts trying to address security and privacy concerns while meeting the requirements in smart grid. For example, Kamto et al. in [17] used encryption to prevent unauthorized access energy profiles. Baumeister in [6] implemented a public key infrastructure in smart grid, which meets most requirements of smart grid, such as scalability and flexibility. In [11], the authors proposed a lightweight Diffe–Hellman authentication mechanism, with Diffe–Hellman key exchange and hash-based authentication technique. Metke et al. in [21] established a secure communication channel if the smart meters are based on trusted computing platform. Nevertheless, the cryptographic solutions can only keep data protected during transmissions, but not for the cases that the adversary compromises the utility providers' servers, or the utility providers themselves are not trustworthy. Under the assumption that the utility provider is semi-honest, i.e., *honest-but-curious* (e.g., [4, 26]), in this book, we propose to allow customers to add distributed differential noises to the measured data before the smart meters send it to the utility provider. Based on the aggregated "noisy" but statistically correct data, we let the utility provider employ data-driven approach to characterize the uncertainty of customers' power demand, match the demand with the supply, and try to minimize the cost of energy generation. We show that the proposed scheme can effectively reduce the power generation cost of the utility provider while preserving the customers' differential privacy in smart grid.

1.4   Information-Centric Networking in CPS

As the rapid increasing of content demands in the Internet, new information-centric networking (ICN) design is motivated to be developed in the future Internet for improved delivery efficiency, content scalability and availability [2, 20, 24]. In addition, ICN architectures are based on named content, which is radically different from the traditional host-centric paradigm based on named hosts [2]. In this new ICN architecture, with the deployment of in-network storage for caching in the access points (AP), it is efficient to offload the tremendous increasing amount of content. In 2017, Cisco highlights that the video traffic has already reached 73% of all the Internet traffic in 2016, and it is estimated to be increased to 82% by 2021 [7]. Inspired by the fact of the speedy growth of the demand for video contents, build-in caching features are supposed to be applied widely in the ICN.

Since the content provider (CP) aims to provide high quality of service (QoS) to the users, the storage for caching in APs plays an important role in reducing the network congestion and backhaul load. As the cache-enabled APs such as base stations are required to cooperate with the CP, it is necessary to find an approach to offering the economic incentives for the contributions and efficiently allocating the resources. As a result, the CP is able to cache the popular data objects with the cooperation of the cache-enabled APs by offering appreciable economic incentives. However, to the best of our knowledge, in previous works they all use the Zipf discrete distribution [1, 14, 20] to represent the content popularity in Internet. As the universal Zipf distribution may not perfectly capture the statistical features of content popularity in various geographical locations in ICN, in our work, we employ data-driven methodology to predict the content popularity from the collected data of local CP users without premise on the content popularity distribution.

While it is beneficial to ICN users with high QoS, it may compromise the users' privacy. To predict the content popularity, the CP aggregates the preferred content information from certain users. However, this aggregation process may elevate risks of privacy leakage. As the user's content preferences may include some sensitive information, these kind of sensitive personal information could be sold as a commodity for commercial uses. For example, because of the disclosure of the private content preference data, users may receive a plenty of spam or fraud emails or phone calls. Therefore, it is necessary to pay attention on protecting on users' private content preferences. For instance, in [24], the authors propose a tag forgery based privacy-enhancing technology to protect the users' interests and preferences in social-tagging systems. In [23], the authors design a tag suppression scheme based on data perturbation to protect end-user privacy in collaborative tagging services.

In our work, in order to address those issues above, we propose a scheme that the CP offloads popular contents into several storage for caching of APs according to the noisy content preference data from users. Therefore, the users' privacy is preserved and the problem of high backhaul load is resolved. Briefly, the CP exploits the data-driven methodology to predict the content popularity distribution according to the collected noisy content preference data from the users and stimulates the APs with economic incentives to lease their storage for caching popular contents. Consequently, we formulates a revenue maximization problem based on the description above and demonstrates that the CP revenue can be effectively optimized, while preserving the customers' local differential privacy in the ICN.

1.5　　Colocation Data Centers in CPS

Demand response is identified as a high-prioritized area, with its potential to reduce up to 20% of the total peak electricity demand of the U.S. [10]. Data center demand response is transforming data centers' huge energy consumption from a negative to a valuable opportunity. Data centers are key participants in emergency demand response (EDR), where the grid coordinates large electricity consumers for reducing their consumption during emergency situations to prevent major economic losses. In the existing literature, many works focused on emergency demand response of data centers [3, 12, 13, 18]. While existing literature concentrates on owner-operated data centers (e.g., Google) where operators have full control over the servers, this work studies EDR in multi-tenant colocation data centers (e.g., Equinix) where servers are owned and managed by individual tenants and which are better targets of EDR.

Colocation, an integral segment of the data center industry, provides unique data center solutions to many industry sectors, including leading IT firms and cloud providers. Due to the increasing global colocation market and its part in huge energy consumption of data centers [22], demand response in colocation data centers has been receiving significant attention. The first study on colocation economic demand response is iCODE [25], which relies on the tenants' best-effort reduction, but cannot assure the truthfulness of strategic tenants. In terms of emergency demand response, the work in [27] proposes a randomized auction mechanism that can guarantee a 2-approximation of social welfare cost and is approximately truthful. Both are based on a reverse auction where tenants must voluntarily submit bids first, and the operator will decide winning bids as well as the reward amount later. However, tenants at first are not concerned with power reduction, so treating their bids as voluntary tasks can lead to pessimistic results on the number of participating tenants. Hence, it is expected that an upfront incentive by the operator will effectively increase tenant participation. Furthermore, in such auctions, tenants need to first calculate and disclose complex bids (e.g., cost functions), which might leak their private information. Existing EDR designs that incentivize tenants' energy reduction can either be gamed by strategic tenants or untrustworthy colocation operators for illegal gains. Colocation operators can be semi-honest and disclosing such individual private information of tenants to operators can pose serious privacy concerns to the tenants. This may preclude tenants' participation in EDR. Not many works in the existing literature addressed the privacy concerns during emergency demand response in data centers.

To address the privacy concerns in colocation data centers, we propose a privacy preserving homomorphic encryption for aggregation to encrypt tenants' private information in this work. In contrast to the existing works, we take a forward-mechanism approach, where the energy reduction and reward allocation rules are announced in advance in order to align tenants' interests to the socially optimal performance. Existing literature for incentivizing tenants' energy reduction in colocation data centers are based on the reverse auction. This work is based on the clock auction, where price/energy is adjusted based on the demand over the course of auction that is essentially required for effective aggregation of energy during EDR. We propose a descending clock auction based emergency demand response where encrypted tenants' cumulative energy reduction is aggregated to meet the EDR, where operator can only know the aggregate of the tenants' values or bids but not their individual private values or confidential information submitted to meet the EDR. We evaluate the privacy and performance of this scheme by formulating descending clock auction, in which the amount of energy/price the tenants are willing to reduce for a given price/energy to meet EDR is protected.

In the rest of this book, we first introduce the techniques we employed in our works (differential privacy, data-driven methodology and descending clock auction accordingly) for various CPS in Chap. 2. After that, we jointly consider data analysis and privacy issues for various CPS applications, formulate optimization, provide feasible solutions, and show the effectiveness of the proposed works in cognitive radio networks, smart grids, information-centric networks and colocation data centers from Chaps. 3 to 6.

References

1. L.A. Adamic, B.A. Huberman, Zipf's law and the internet. Glottometrics **3**(1), 143–150 (2002)
2. B. Ahlgren, C. Dannewitz, C. Imbrenda, D. Kutscher, B. Ohlman, A survey of information-centric networking. IEEE Commun. Mag. **50**(7), 26–36 (2012)
3. D. Aikema, R. Simmonds, H. Zareipour, Data centres in the ancillary services market, in *International Green Computing Conference (IGCC)*, San Jose, CA (2012), pp. 1–10
4. M.R. Asghar, G. Russello, B. Crispo, M. Ion, Supporting complex queries and access policies for multi-user encrypted databases, in *Proceedings of the ACM Cloud Computing Security Workshop, Co-located with CCS*, Berlin (2013)
5. B. Bahrak, S. Bhattarai, A. Ullah, J.-M. Park, J. Reed, D. Gurney, Protecting the primary users operational privacy in spectrum sharing, in *IEEE International Symposium on Dynamic Spectrum Access Networks*, Mclean, VA (2014)
6. T. Baumeister, Adapting pki for the smart grid, in *IEEE International Conference on Smart Grid Communications (SmartGridComm)*, Brussels (2011)
7. Cisco visual networking index: forecast and methodology, 2016–2021. Cisco (2017)
8. L. Duan, J. Huang, B. Shou, Cognitive mobile virtual network operator: investment and pricing with supply uncertainty, in *Proceedings of IEEE Conference on Computer Communications, INFOCOM*, San Diego, CA (2010)
9. C. Dwork, Differential privacy: a survey of results, in *International Conference on Theory and Applications of Models of Computation* (Springer, Berlin, 2008), pp. 1–19
10. Federal Energy Regulatory Commission, A national assessment of demand response potential (2009)
11. M.M. Fouda, Z.M. Fadlullah, N. Kato, R. Lu, X.S. Shen, A lightweight message authentication scheme for smart grid communications. IEEE Trans. Smart Grid **2**(4), 675–685 (2011)
12. M. Ghamkhari, H. Mohsenian-Rad, Data centers to offer ancillary services, in *IEEE Third International Conference on Smart Grid Communications (SmartGridComm)*, Taiwan (2012), pp. 436–441
13. G. Ghatikar, Demand response opportunities and enabling technologies for data centers: findings from field studies (2014)
14. M. Hajimirsadeghi, N.B. Mandayam, A. Reznik, Joint caching and pricing strategies for popular content in information centric networks. IEEE J. Sel. Areas Commun. **35**(3), 654–667 (2017)

15. Q. Huang, Y. Tao, F. Wu, Spring: a strategy-proof and privacy preserving spectrum auction mechanism, in *Proceeding of IEEE International Conference on Computer Communications (INFOCOM)*, Turian (2013)
16. X. Jin, Y. Zhang, Privacy-preserving crowdsourced spectrum sensing, in *Proceeding of the IEEE International Conference on Computer Communications (INFOCOM)* (2016), pp. 1–9
17. J. Kamto, L. Qian, J. Fuller, J. Attia, Y. Qian, Key distribution and management for power aggregation and accountability in advance metering infrastructure, in *IEEE Third International Conference on Smart Grid Communications (SmartGridComm)*, Tainan (2012)
18. Z. Liu, I. Liu, S. Low, A. Wierman, Pricing data center demand response. ACM SIGMETRICS Perform. Eval. Rev. **42**(1), 111–123 (2014)
19. X. Lou, R. Tan, D.K. Yau, P. Cheng, Cost of differential privacy in demand reporting for smart grid economic dispatch, in *Proceeding of the IEEE International Conference on Communications (ICC'17)*, Paris (2017)
20. M. Mangili, F. Martignon, S. Paris, A. Capone, Bandwidth and cache leasing in wireless information-centric networks: a game-theoretic study. IEEE Trans. Veh. Technol. **66**(1), 679–695 (2017)
21. A.R. Metke, R.L. Ekl, Security technology for smart grid networks. IEEE Trans. Smart Grid **1**(1), 99–107 (2010)
22. NRDC, Scaling up energy efficiency across the data center industry: evaluating key drivers and barriers. NRDC Issue Paper (2014)
23. J. Parra-Arnau, A. Perego, E. Ferrari, J. Forne, D. Rebollo-Monedero, Privacy-preserving enhanced collaborative tagging. IEEE Trans. Knowl. Data Eng. **26**(1), 180–193 (2014)
24. S. Puglisi, J. Parra-Arnau, J. Forné, D. Rebollo-Monedero, On content-based recommendation and user privacy in social-tagging systems. Comput. Stand. Interfaces **41**, 17–27 (2015)
25. S. Ren, M.A. Islam, Colocation demand response: Why do i turn off my servers?, in *11th International Conference on Autonomic Computing (ICAC 14)*, Philadelphia (2014), pp. 201–208
26. O. Vukovic, G. Dan, R.B. Bobba, Confidentiality-preserving obfuscation for cloud-based power system contingency analysis, in *IEEE International Conference on Smart Grid Communications (SmartGridComm)*, Vancouver (2013)
27. L. Zhang, S. Ren, C. Wu, and Z. Li, A truthful incentive mechanism for emergency demand response in colocation data centers, in *IEEE Conference on Computer Communications (INFOCOM)*, Hong Kong (2015), pp. 2632–2640
28. R. Zhu, Z. Li, F. Wu, K. Shin, G. Chen, Differentially private spectrum auction with approximate revenue maximization, in *Proceedings of ACM International Symposium on Mobile Ad Hoc Networking and Computing, ACM MobiHoc* (2014), pp. 185–194

Preliminaries

2

Abstract

In this chapter, we briefly introduce the differential privacy technique and its variants to effectively protect the participants' privacy in different CPS. The concept of *differential privacy* was first proposed by Dwork (Differential privacy: a survey of results. In: International conference on theory and applications of models of computation, Xi'an, 2008) which specifies that any individual has a very small influence on the (distribution of the) outcome of the computation. Differential privacy (DP) aims to exploit the statistical information without disclosure of the data providers' privacy. Differential privacy is a formal definition of data privacy, which ensures that any sequence of output from data set (e.g., responses to queries) is "essentially" equally likely to occur, no matter any individual is present or absent (Dwork et al., Found Trends Theor Comput Sci 9(3–4):211–407, 2014; Baranov et al., Am Econ J Microecon 9(3):1–27, 2017; Jin and Zhang, Privacy-preserving crowdsourced spectrum sensing. In: Proceeding of the IEEE international conference on computer communications (INFOCOM), pp. 1–9, 2016). In this chapter, we illustrate three variants of differential privacy, centralized different privacy, distributed differential privacy and local differential privacy that are applied to various CPS applications described in the book.

2.1 Differential Privacy

2.1.1 Centralized Differential Privacy

In the standard (or centralized) DP setting, each user sends raw data to the database, who obtains the true distribution, adds noise, and then publishes the result. In this setting, the aggregator is trusted to not reveal the raw data and is trusted to

© The Author(s), under exclusive license to Springer Nature Switzerland AG 2019   11
M. Pan et al., *Big Data Privacy Preservation for Cyber-Physical Systems*,
SpringerBriefs in Electrical and Computer Engineering,
https://doi.org/10.1007/978-3-030-13370-2_2

handle the raw data correctly [6]. DP keeps the characteristic of the whole data set, and preserves information privacy of each individual. The definition of ϵ-DP is as follows.

Definition 1 (Differential Privacy) Let $\mathscr{A}$ denote a randomized algorithm. We take the output as $\mathbf{r}$ and the input as x, i.e., $\mathscr{A}(x) = \mathbf{r}$. For all $x, x' \subseteq \mathcal{N}^{|\mathcal{X}|}$ satisfied $||x - x'|| \leq 1$,

$$log \frac{Pr(\mathbf{r}|x)}{Pr(\mathbf{r}|x')} \leq \epsilon. \tag{2.1}$$

Then we call $\mathscr{A}$ is ϵ-DP. The parameter ϵ represents the degree of DP offered, which is the upper bond of the differences between true $\mathscr{A}(x)$ and $\mathscr{A}(x')$. A smaller value of ϵ implies the stronger privacy guarantee and perturbation noise, and a larger value of ϵ implies a weaker privacy guarantee while having higher data utility.

Definition 2 (l_1-Sensitivity of a Function f) The l_1-sensitivity of a function $f : \mathbb{N}^{|\mathcal{X}|} \to \mathbb{R}^k$ is

$$\Delta f = \max_{\substack{x,x' \in \mathbb{N}^{|\mathcal{X}|}, \\ ||x-x'||_1=1}} ||f(x) - f(x')||_1. \tag{2.2}$$

The l_1 sensitivity of a function f captures the magnitude by which a single individual's data can change the function f in the worst case, and therefore, intuitively, the uncertainty in the response that we must introduce in order to hide the participation of a single individual [9].

Definition 3 (The Laplace Distribution and the Laplace Mechanism) The Laplace Distribution (centered at 0) with scale b is the PDF (Probability Density of Function) is:

$$Lap(x|b) = \frac{1}{2b}\exp(-\frac{|x|}{b}). \tag{2.3}$$

In the rest of this book, we will write Lap(b) simply to denote a random variable $X \sim$ Lap(b). The *Laplace Mechanism* simply computes f, and perturb each coordinate with noise drawn from the Laplace distribution. The scale of the noise will be calibrated to the sensitivity of f. The Laplace Mechanism $\mathscr{A}_L$ is defined as

$$\mathscr{A}_L(x, f(\cdot), \epsilon) = f(x) + (Y_1, \cdots, Y_k),$$

where Y_i are i.i.d random variables drawn from Lap($\Delta f/\epsilon$). The proof of Laplace mechanism reserves ϵ-DP is shown in [9].

2.1.2   Distributed Differential Privacy

As introduced in the previous section, the centralized differential privacy setting has a strong assumption that a trustworthy third-party database or data aggregator is required to apply the randomized algorithm on the exact data of data providers. In reality, people may evade to provide data to a survey including sensitive questions. In such a situation that the data providers trust no one even the data collectors but only themselves, secure multi-party computation and homomorphic encryption are suitable to involve in the differential privacy definition. In [19], the authors propose a private stream aggregation algorithm which guarantees distributed differential privacy of individual user, while the aggregator can only get the statistical results, but not learn any other unintended information from users. We assume $\mathbf{x} = \{x_1, \cdots, x_n\}$ denotes the vector of users' energy demand, function f represents the desired statistics of utility provider and $\mathcal{O}$ indicates the output range of function f. Then, the definition of *distributed differential privacy* (DDP) is shown as follows.

Definition 2.1 With a privacy confidence parameter $\epsilon > 0$ and $0 \leq \delta < 1$, a randomized algorithm $\mathcal{A}$ satisfies (ϵ, δ)-distributed differential privacy with respect to the function f, for any subset $X \subseteq \mathcal{O}$, when given two neighbor vectors $\mathbf{x}, \mathbf{x}'$ [19]:

$$Pr[f(\mathcal{A}(\mathbf{x})) \in X] \leq e^\epsilon \cdot Pr[f(\mathcal{A}(\mathbf{x}')) \in X] + \delta.$$

The two neighbor vectors $\mathbf{x}, \mathbf{x}'$ are supposed to have only one element different. The privacy confidence parameter ϵ controls the privacy preservation level. With smaller ϵ, it is less possible to distinguish the outputs of the randomized algorithm $\mathcal{A}$ with two different inputs, which means the privacy protection is stronger.

In the setup of the DDP algorithm, based on the homomorphic encryption scheme, each customer gets a private key sk_j to encrypt the demand with distributed differential private noise and the utility provider also gets the key sk_0 to decrypt the statistics when receiving all of the cipher texts from customers. Because the utility provider can only learn the noisy statistic, each customer would add less noise to the demand data, if the randomization of the desired statistic $f(\mathcal{A}(\mathbf{x}))$ is big enough. As the customer's data is in a discrete group, in DDP, a symmetric geometric distribution is exploited to guarantee the satisfaction of differential privacy. The probability mass function of symmetric geometric distribution is $\text{Geom}(\alpha) = \frac{\alpha-1}{\alpha+1}\alpha^{-|j|}$. When α is set to $e^{\epsilon/\Delta}$ and the noise comes from this symmetric geometric distribution, the randomized algorithm $\mathcal{A}$ is able to achieve differential privacy [19]. In our scenario, since we are focusing on the summation of all the residential users' demand, the sensitivity Δ is supposed to be the maximum energy demand of an individual customer. After achieving the differential privacy, each customer sends the noisy energy demand data encrypted with private key sk_j to the utility provider. With the aggregation of all the encrypted noisy energy demand data, the utility provider is able to sufficiently decrypt the summation of the data [22] with the key sk_0. Therefore, the utility provider is supposed to learn only the summation of users' energy demand and no information from each user. With the DDP algorithm, the utility provider is able to get the summation demand during a time period $d = f(\mathscr{A}(\mathbf{s}))$, which is a noisy version of $f(\mathbf{s})$, at the meanwhile, the differential privacy of each individual customer is guaranteed.

2.1.3 Local Differential Privacy

Differential privacy [8] is used to obtain the statistical information of databases without disclosure of the data providers' privacy. Intuitively, given two databases, which have only one element different from each other, as the inputs of a randomization algorithm, the outputs are not distinguishable. However, there must exist a trustworthy database or data aggregator when applying the centralized differential privacy. In local differential privacy, it assumes that the service database is *honest-but-curious*, which means the privacy leakage possibility increases. Therefore, local privacy setting is suitable in the situation that the data providers trust no one except themselves. The Warner's random response model [21] is one of the oldest local privacy model applied in survey sampling. If there are two answers of one question, the data provider will reply truly with probability of p and falsely with probability of $1 - p$. Combining local privacy and differential privacy, the definition of *local differential privacy* (LDP) is shown as follows.

Definition 2.2 With a privacy confidence parameter $\epsilon \geq 0$, a randomized algorithm $\mathscr{A}$ satisfies ϵ-local differential privacy, when given two inputs x and x' [7]:

$$\frac{Pr[\mathscr{A}(x) = z]}{Pr[\mathscr{A}(x') = z]} \leq e^{\epsilon},$$

where z is the secure view of the input.

Therefore, with a specific output z from the randomized algorithm $\mathscr{A}$, it is not able to determine or can infer with negligible probability whether the input is x or x'. Additionally, the privacy confidence parameter ϵ controls the privacy preservation level, which means there is more possibility to distinguish the outputs of the randomized algorithm $\mathscr{A}$ with two different inputs with a higher value of ϵ. In other words, smaller ϵ means higher privacy preservation level.

To perform a LDP mechanism, it contains several steps. First, the true data is encoded locally into a vector or a number. Next, the encoded data is randomized by a specific function. At last, the processed data will be sent to the data aggregator or database. Among the three steps, the combination of the first two steps is the randomized algorithm $\mathscr{A}$ in the definition, which is finished locally and supposed to satisfy ϵ-LDP. In [20], the authors have introduced an optimized LDP protocol, named optimal local hashing (OLH), which can offer higher accuracy of frequency estimation with lower communication cost.

In the encoding step of OLH, the input, denoted as $r_u \in [1, F]$, is first encoded with the hash function H, which can hash the input value into $[g]$ ($g > 2$), uniformly chosen from a universe hash function family $\mathbb{H}$. The output of first step is represented $r_u^H = Encode(r_u) = \langle H, r_u \rangle$. In the next perturbation step, r_u^H is perturbed into $r_u' = Perturb(r_u^H)$, with the probability shown as follows:

$$\forall_{k \in [g]} Pr[r_u' = k] = \begin{cases} p = \frac{e^\epsilon}{e^\epsilon + g - 1}, \text{ when } r_u^H = k, \\[2mm] q = \frac{1}{e^\epsilon + g - 1}, \text{ when } r_u^H \neq k. \end{cases} \quad (2.4)$$

This local hashing protocol including encoding and perturbation is satisfactory to ϵ-LDP (the detailed proof shown in [20]). With $g = e^\epsilon + 1$, it can receive the optimal variance of the aggregation results, which is used to estimate the frequency of each input value reported. Therefore, after employing the OLH protocol, one user's input r_u will be r_u' within the domain of g. During the aggregation process, the CP can estimate the frequency of each value r_u' in the range F occurs from the following equation,

$$r_u(f) = \frac{\sum_{u=1}^{U} I_f(r_u') - Nq^*}{p^* - q^*}, \quad (2.5)$$

where $\sum_{u=1}^{U} I_f(r_u')$ is the counts of occurrence of each value r_u' in the range F, p^* is equal to p from (2.4) is and $q^* = \frac{1}{g}$ (detailed description is in [20]).

2.2 Big Data Analysis: Data-Driven Methodology Preliminaries

2.2.1 ζ-Structure Probability Metrics

We use a ζ-structure probability metric, which is a distribution distance measurement proposed in [5, 12] to quantify the distance of distributions. Specifically, a predefined distance measure $d(\mathbb{P}_0, \mathbb{P})$ is constructed on confidence set $\mathscr{D}$, where $\mathbb{P}$ is the true distribution and $\mathbb{P}_0$ is the reference distribution conducted from historical data. The distance d_ζ and confidence set $\mathscr{D}$ can be defined as follows,

$$\mathscr{D} = \{\mathbb{P} : d_\zeta(\mathbb{P}_0, \mathbb{P}) \leq \theta\}, \quad (2.6)$$

$$d_\zeta(\mathbb{P}_0, \mathbb{P}) = \sup_{h \in \mathscr{H}} \left| \int_\Omega h d\mathbb{P}_0 - \int_\Omega h d\mathbb{P} \right|. \quad (2.7)$$

Here, $d_\zeta(\cdot, \cdot)$ represents the distance under ζ structure probability metric, θ denotes the tolerance, and $\mathscr{H}$ is a family of real-valued bounded measurable functions on Ω (the sample space on ξ). Tolerance θ is correlated to data size Q, i.e., the size of historical data. It can be easily inferred that the more demand samples that the STED can collect, the tighter $\mathscr{D}$ would be, and the closer ambiguous distribution $\mathbb{P}_0$ would be to $\mathbb{P}$. More details of ζ-structure probability metric is introduced in the next section.

2.2.2 Converge Rate Under ζ-Structure Probability Metrics

Three ζ-probability metrics are employed to solve the proposed problem, which are derived as follows. We define $\rho(x, y)$ as the distance between two variables x and y. $\mathbb{P} = \mathscr{L}(x)$ as random variables x following distribution $\mathbb{P}$.

- **Kantorovich metric**: denoted as $d_K(\mathbb{P}_0, \mathbb{P})$, $\mathscr{H} = \{h : ||h||_L \leq 1\}$, where $||h||_L := \sup\{h(x) - h(y)/\rho(x, y) : x \neq y \text{ in } \Omega\}$. By the Kantorovich–Rubinstein theorem, the Kantorovich metric is equivalent to the Wasserstein metric. In particular, when $\Omega = R$, let d_w denote the Wasserstein metric, then

$$d_w(\mathbb{P}_0, \mathbb{P}) = \int_{-\infty}^{+\infty} |F(x) - G(x)| dx, \qquad (2.8)$$

where F and G are the distribution function derived from $\mathbb{P}_0$ and $\mathbb{P}$ respectively, which is demonstrated in Fig. 2.1.
- **Fortet–Mourier metric**: denoted as $d_{FM}(\mathbb{P}_0, \mathbb{P})$, $\mathscr{H} = \{h : ||h||_C \leq 1\}$, where $||h||_C := \sup\{h(x) - h(y)/c(x, y) : x \neq y \text{ in } \Omega\}$ and $c(x, y) = \rho(x, y) \max\{1, \rho(x, a)^{p-1}, \rho(y, a)^{p-1}\}$ for some $p \geq 1$ and $a \in \Omega$. Note that when $p = 1$, Fortet–Mourier metric is the same as Kantorovich metric. The Fortet–mourier metric is usually utilized as a generalization of Kantorovich metric, with the application on mass transportation problems.
- **Uniform metric**: denoted as $d_U(\mathbb{P}_0, \mathbb{P})$, $\mathscr{H} = \{I_{(-\infty, t]}, t \in R^n\}$. According to the definition, we have $d_U(\mathbb{P}_0, \mathbb{P}) = \sup_t |\mathbb{P}_0(x \leq t), \mathbb{P}(x \leq t)|$. It is illustrated in Fig. 2.2, where F and G are the distribution functions derived from $\mathbb{P}$ and $\mathbb{P}_0$, respectively.

Fig. 2.1 Wasserstein metrics (one-dimensional case)

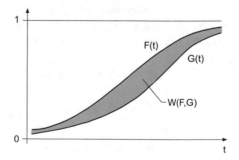

Fig. 2.2 Uniform metric

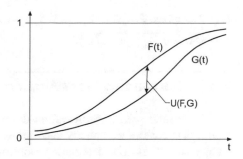

From the definition of metrics and relationships between metrics under ζ-structure, we can derive the convergence property and convergence rate accordingly. For the uniform metric, the convergence rate can be derived from the Dvoretzky–Kiefer–Wolfowitz inequality [4, 11, 18]:

Proposition 1 *The convergence rate of the uniform metric for a single dimension case is (i.e., $n = 1$),*

$$\mathbb{P}(d_U(\mathbb{P}_0, \mathbb{P}) \leq \theta) \geq 1 - \exp\left(-\frac{\theta^2 Q}{2}\right). \tag{2.9}$$

In [23], the converge rate of the Kantorovich metric is shown below:

Proposition 2 *For a general dimension case (i.e., $n \geq 1$),*

$$\mathbb{P}(d_K(\mathbb{P}_0, \mathbb{P}) \leq \theta) \geq 1 - \exp\left(-\frac{\theta^2 Q}{2\varnothing^2}\right). \tag{2.10}$$

Therefore we have $(\mathbb{P}(\mathbb{P}_0, \mathbb{P}) \leq \theta) \geq 1 - \exp(-\frac{\theta^2}{2\varnothing^2}Q) = \eta$, and $\theta = \varnothing\sqrt{2log(1/(1-\eta))/Q}$.

The relationship among metrics is represented as $d_{FM}(\mathbb{P}_0, \mathbb{P}) \leq \Lambda \cdot d_K(\mathbb{P}_0, \mathbb{P})$, where $\Lambda = \max\{1, \varnothing^{p-1}\}$ and $\varnothing$ is the diameter of Ω. From the relation between the Fortet–Mourier metric and Kantorovich metric with Proposition 2, we can easily derive the convergence rate of other metrics.

Corollary 1 *For a general dimension (i.e., $n \geq 1$), we have*

$$\mathbb{P}(d_{FM}(\mathbb{P}_0, \mathbb{P}) \leq \theta) \geq 1 - \exp\left(-\frac{\theta^2 Q}{2\varnothing^2\Lambda^2}\right). \tag{2.11}$$

With the convergence rate in (2.9)–(2.11), we can calculate the tolerance θ accordingly. For instance, in the Kantorovich metric, we assume the confidence level

is η. Therefore, $\mathbb{P}(d_u(\mathbb{P}_0, \mathbb{P}) \leq \theta)) \geq 1 - \exp(-\frac{\theta^2}{2\varnothing^2} Q) = \eta$ according to (2.9), and
$\theta = \varnothing\sqrt{2log(1/(1 - \eta)/Q)}$.

2.3   Descending Clock Auction

For differential privacy preservation during emergency demand response in colocation data centers we employ descending clock auction for incentivizing tenants' energy reduction. The descending (multi-item) clock auction (DCA) [1, 16] is a mechanism for buying items from multiple potential sellers [2, 3, 13, 14, 17]. We consider price DCA to be effective, since EDR is required during emergency situations to prevent major economic losses, providing incentives starting at higher reserve price like DCA may interest the tenants in immediately participating for EDR. Incentives through ascending clock auction that start at low price and increment over course of auction might keep the operator waiting, since the tenants aspire for higher incentives. Hence, we employ DCA in the EDR scenario, where we consider the auctioneer/operator as a buyer willing to pay specific price for an item (i.e., a unit of energy say per kWh) and the tenants/bidders as sellers who submit the number of items (i.e., amount of energy in terms of kWh). We have a set of $\mathcal{N}$ bidders, where each tenant $i \in \mathcal{N}$ has specific amount of energy e_i and decides to decrease it or not based on the offer price. In the DCA, the bidder-specific prices are initialized at reserve prices and then decremented over the course of the auction. In each round, the auctioneer decrements the offer prices to the bidders, who might accept or decline the offers. Accepting means that the bidder is willing to reduce their energy consumption at that price. Rejecting means that the bidder has no interest in reducing the energy consumption. This process is repeated until the auctioneer's amount of energy reduction would become infeasible if additional bidders were to reject any new (lower) offers or when the target amount of energy reduction is met under the specified budget. At that point the auction ends and the current prices are paid. Similar to the price descending clock auction, we consider energy descending clock auction where tenants submit the reward price to the operator they are willing to receive for reducing their energy. Since EDR is required during emergency situations to prevent major economic losses, energy descending clock auction helps to meet the energy reduction target in less time compared to the ascending case that keeps the operator waiting to meet the energy reduction target. As the tenants can quote higher prices in such a case, we provide an architecture considering multiple operators and tenants to determine the clearing price/equilibrium price [10, 15] guaranteeing incentive compatibility. Because of ever-decreasing bids in descending clock auctions, buyers must act decisively to name their energy/price or may risk losing to a lower offer.

References

1. L.M. Ausubel, P. Cramton, Auctioning many divisible goods. J. Eur. Econ. Assoc. **2**(2–3), 480–493 (2004)
2. O. Baranov, C. Aperjis, L.M. Ausubel, T. Morrill, Efficient procurement auctions with increasing returns. Am. Econ. J. Microecon. **9**(3), 1–27 (2017)
3. D. Bergemann, B.A. Brooks, S. Morris, Optimal auction design in a common value model (2016). http://dx.doi.org/10.2139/ssrn.2886453
4. F. Bolley, C. Villani, Weighted csiszar-kullback-pinsker inequalities and applications to transportation inequalities. Ann. Fac. Sci. Toulouse **14**, 331–352 (2005)
5. G.C. Calafiore, Ambiguous risk measures and optimal robust portfolios. SIAM J. Optim. **18**(3), 853–877 (2007)
6. G. Cormode, S. Jha, T. Kulkarni, Privacy at scale: local differential privacy in practice, in *2018 ACM SIGMOD/PODS International Conference on Management of Data*, Houston, TX (2018)
7. J.C. Duchi, M.I. Jordan, M.J. Wainwright, Local privacy and statistical minimax rates, in *2013 IEEE 54th Annual Symposium on Foundations of Computer Science (FOCS)*, Berkeley, CA (2013)
8. C. Dwork, Differential privacy: a survey of results, in *International Conference on Theory and Applications of Models of Computation*, Xi'an (2008)
9. C. Dwork, A. Roth, et al., The algorithmic foundations of differential privacy. Found. Trends Theor. Comput. Sci. **9**(3–4), 211–407 (2014)
10. S. Gjerstad, J. Dickhaut, Price formation in double auctions. Games Econ. Behav. **22**(1), 1–29 (1998)
11. J.M. Hammersley, D.C. Handscomb, *Monte Carlo Methods* (Methuen, London, 1964)
12. D. Klabjan, D. Simchi-Levi, M. Song, Robust stochastic lot-sizing by means of histograms. Prod. Oper. Manag. **22**(3), 691–710 (2013)
13. J. Levin, A. Skrzypacz, Are dynamic vickrey auctions practical?: properties of the combinatorial clock auction. Technical report, National Bureau of Economic Research (2014)
14. D. Liu, A. Bagh, New privacy-preserving ascending auction for assignment problems (2016). http://dx.doi.org/10.2139/ssrn.2883867
15. R.P. McAfee, A dominant strategy double auction. J. Econ. Theory **56**(2), 434–450 (1992)
16. T.-D. Nguyen, T. Sandholm, Optimizing prices in descending clock auctions, in *Proceedings of the Fifteenth ACM Conference on Economics and Computation*, Stanford (2014), pp. 93–110
17. R. Poudineh, T. Jamasb, Distributed generation, storage, demand response and energy efficiency as alternatives to grid capacity enhancement. Elsevier Energy Policy **67**, 222–231 (2014)
18. U. Schmock, Large deviations techniques and applications. J. Am. Stat. Assoc. **95**(452), 1380–1381 (2000)
19. E. Shi, H. Chan, E. Rieffel, R. Chow, D. Song, Privacy-preserving aggregation of time-series data, in *Annual Network & Distributed System Security Symposium (NDSS)*, San Diego, CA (2011)
20. T. Wang, J. Blocki, N. Li, S. Jha, Locally differentially private protocols for frequency estimation, in *Proceedings of the 26th USENIX Security Symposium*, Vancouver, BC (2017)
21. S.L. Warner, Randomized response: a survey technique for eliminating evasive answer bias. J. Am. Stat. Assoc. **60**(309), 63–69 (1965)

22. Z. Yang, S. Zhong, R.N. Wright, Privacy-preserving classification of customer data without loss of accuracy, in *Proceedings of the 2005 SIAM International Conference on Data Mining*, Newport Beach, CA (2005)
23. C. Zhao, Y. Guan, Data-driven risk-averse two-stage stochastic program with ζ-structure probability metrics. Optim. Online **2**, 9 (2015)

Spectrum Trading with Secondary Users' Privacy Preservation

3

Abstract

As described in Sect. 1.2, spectrum trading benefits both SUs and PUs, while it poses great challenges to maximize PUs' revenue, since SUs' demands are uncertain and individual SU's traffic portfolio contains private information. In this chapter, we propose a data-driven spectrum trading scheme which maximizes PUs' revenue and preserves SUs' demand differential privacy. Briefly, we introduce a novel network architecture consisting of the PSP, the SSP and the STED. Under the proposed architecture, PSP aggregates available spectrum from PUs, and sells the spectrum to SSP at fixed wholesale price, directly to SUs at spot price, or both. The PSP has to accurately estimate SUs' demands. To estimate SUs' demand, the STED exploits data-driven approach to choose sampled SUs to construct the reference distribution of SUs' demands, and utilizes reference distribution to estimate the demand distribution of all SUs. Moreover, the STED adds noises to preserve the demand differential privacy of sampled SUs before it answers the demand estimation queries from the PSP. With the estimated SUs' demand, we formulate the revenue maximization problem into a risk-averse optimization, develop feasible solutions, and verify its effectiveness through both theoretical proof and simulations.

3.1 System Description and 3DPP Outline

3.1.1 System Model and Adversary Model

Our proposed spectrum trading market consists of the PSP, the STED, the SSP, and $\mathcal{N} = \{1, 2, \cdots, i, \cdots, N\}$ SUs as shown in Fig. 3.1. As introduced in Sect. 1.2, the PSP and the SSP are entities similar to MVNOs, and the STED is a trustworthy database server for SUs, which can collect the traffic demand information from SUs,

© The Author(s), under exclusive license to Springer Nature Switzerland AG 2019   21
M. Pan et al., *Big Data Privacy Preservation for Cyber-Physical Systems*,
SpringerBriefs in Electrical and Computer Engineering,
https://doi.org/10.1007/978-3-030-13370-2_3

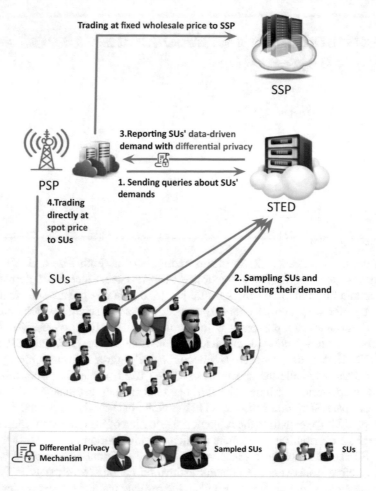

Fig. 3.1 The spectrum trading procedure of 3DPP

and temporarily store it [2, 6, 7]. The PSP is entitled to aggregate vacant spectrum resources from $\mathscr{M} = \{1, 2, \cdots, j, \cdots, M\}$ PUs with unequal sized bandwidth $\mathscr{W} = \{W_1, W_2, \cdots, W_j, \cdots, W_M\}$, and sell those available spectrum bands for monetary gains. Similar to power market/cloud resource market in smart grid/cloud computing systems, the PSP has the following spectrum trading options: (1) selling available spectrum to the SSP at fixed wholesale price, i.e., c; (2) selling available spectrum bands to the SUs directly at spot price, i.e., b; or (3) dividing available spectrum resources and selling to both. Thus, before splitting the spectrum and deciding the selling strategy, the PSP will send queries about SUs' demands to the STED as shown in Fig. 3.1. Due to the large number of SUs within the PSP's coverage, the STED will sample some SUs, build up a reference traffic demand distribution of SUs, and respond to the PSP's queries.

The adversaries could be the dishonest PSP or eavesdropping attackers, who are always monitoring the information exchange between the PSP and the STED. As shown in Fig. 3.1, without enforcing any privacy preserving schemes, the adversaries can easily learn the sampled SUs' traffic demand profiles. That may help the adversaries make some illegal monetary gains, or even launch jamming attacks on some valuable services of chosen SUs. It also makes the SUs reluctant to participate in spectrum trading.

3.1.2   3DPP Outline

To preserve the sampled SUs' DP, it takes four steps for the PSP to sell the available spectrum to SUs at spot price b as shown in Fig. 3.1. Firstly, the PSP sends queries about SUs' demands to the STED. Secondly, STED samples some SUs, and constructs a reference demand distribution $\mathbb{P}_0$ from sampled SUs' demands. The STED needs to ensure the uncertainty distance between the reference distribution $\mathbb{P}_0$ and the real traffic demand distribution of all SUs $\mathbb{P}$ is close enough. Thirdly, the STED adds noises drawn from Laplace distribution to $\mathbb{P}_0$, and establishes a SUs' traffic demand reference distribution $\mathbb{P}_0'$, which achieves ϵ-DP [3,5,11]. Meanwhile, the STED needs to guarantee that $\mathbb{P}_0'$ is close enough to $\mathbb{P}$, so that $\mathbb{P}_0'$ satisfies both data-driven and ϵ-DP requirements. Then, the STED responds to the PSP's queries with $\mathbb{P}_0'$. Finally, based on $\mathbb{P}_0'$, the PSP decides how much spectrum needs to be sold to the SUs directly at b, and how much spectrum need to be sold to the SSP at c to maximize its revenue.

Following this spectrum trading procedure, in the next section, we formulate the PSP's revenue maximization problem under data-driven and DP constraints, i.e., 3DPP. In Sect. 3.3, we theoretically prove that $\mathbb{P}_0'$ is close enough to $\mathbb{P}$, which means the proposed 3DPP has data-driven and DP properties. We also develop solutions to 3DPP problem in Sect. 3.3.

3.2   3DPP Problem Formulation

3.2.1   PSP's Revenue Maximization Formulation

Let γ_j be a binary variable indicating if W_j is directly sold to SUs, where $\gamma_j = 1$ if W_j is directly sold to SUs, and 0, otherwise. Thus, the PSP's revenue gained from selling spectrum to the SSP can be written as $\sum_{j=1}^{M} cW_j(1-\gamma_j)$, where $1-\gamma_j$ represents the spectrum sold to the SSP at fixed price c. Besides, let random variable ξ denote the uncertain demands from all SUs, and ξ follows distribution $\mathbb{P}$. Then, $b\left(\min\left(\sum_{j=1}^{M} W_j\gamma_j, \xi\right)\right)$ is the PSP's revenue gained from selling spectrum to SUs directly.[1] Here, due to the uncertainty of SUs' demands, if the spectrum supply from the PSP (i.e., the spectrum bands that the PSP decided to sell to SUs directly) is more than SUs' actual total traffic demand, i.e., $\sum_{j=1}^{M} W_j \gamma_j > \xi$, the revenue for the PSP is $b\xi$. Otherwise, if the spectrum supply from the PSP is less than SUs' actual traffic demand, i.e. $\sum_{j=1}^{M} W_j \gamma_j < \xi$, the revenue for the PSP is $b \sum_{j=1}^{M} W_j \gamma_j$.

Putting those two parts together, the PSP's revenue maximization can be formulated as follows.

$$\max_{\gamma} \quad -\sum_{j=1}^{M} c W_j \gamma_j + \sum_{j=1}^{M} c W_j$$

$$+ b \mathbb{E}_{\mathbb{P}} \Big(\min \big(\sum_{j=1}^{M} W_j \gamma_j, \xi \big) \Big), \tag{3.1}$$

$$\text{s.t.:} \quad \gamma_j \in \{0, 1\}, j = 1, \cdots, M, \tag{3.2}$$

$$\xi = \sum_{i=1}^{N} d_i, i = 1, \cdots, N, \tag{3.3}$$

where γ_j is binary variable, and (3.3) represents the total traffic demand of all SUs.

3.2.2   Data-Driven Based PSP's Revenue Optimization

Given the huge number of SUs within PSP's coverage, the STED cannot collect traffic demand information from every possible SU, i.e., the STED is generally difficult to obtain the true probability distribution of all SUs' demand $\mathbb{P}$. Instead, we allow the STED to collect the traffic demands from a series of sampled SUs, and construct reference demand distribution $\mathbb{P}_0$. For a given set of sampled SU data, it is easy for us to construct a histogram to fit the SUs' traffic demand. For example, we can set N intervals to fit the total traffic demand of sampled SUs in each interval to be $L_1, \cdots, L_n, \cdots, L_N$ with $L = \sum_{n=1}^{N} L_n$. For instance, L_1 is the number of SUs who would like to access spectrum on price \$15/MHz, L_2 is the number of SUs who would like to access spectrum on price \$20/MHz, etc. Based on this, we can construct an reference distribution for the uncertain total traffic demand of all consumers in particular time period of a day as $p_1^0 = L_1/L, \cdots,$ $p_n^0 = L_n/L, \cdots,$ and $p_N^0 = L_N/L$. For simplicity, we let $\mathbb{P}_0 = p_1^0, p_2^0, \cdots, p_N^0$ represent the corresponding reference distribution. Since $\mathbb{P}_0$ may not be 100% represents the unique true SUs' demand distribution $\mathbb{P}$, we employ risk-averse stochastic optimization approaches (RA-SP) allowing distribution ambiguity [9] to

---

[1] In this work, we assume the aggregated spectrum resources can be perfectly split to satisfy SUs' traffic demands.

reformulate the PSP's revenue maximization problem in (3.1). Instead of deriving a true distribution for ξ, this optimization approach derives a confidence set D, and allows the distribution ambiguity to be within set $\mathscr{D}$ with a certain confidence level (e.g., 99%). The data-driven based RA-SP for the PSP's revenue maximization is formulated as follows.

$$\max_{\gamma} \quad -\sum_{j=1}^{M} c W_j \gamma_j + \sum_{j=1}^{M} c W_j$$

$$+ \min_{\mathbb{P}\in\mathscr{D}} b\mathbb{E}_{\mathbb{P}}\Big(\min(\sum_{j=1}^{M} W_j \gamma_j, \xi)\Big), \tag{3.4}$$

s.t.:   constraints (3.2) and (3.3).

3.2.3 3DPP: Data-Driven Based PSP's Revenue Optimization Under ϵ-DP

To protect the sampled SUs' traffic demand profiles, the STED will employ Laplace mechanism to add noises into $\mathbb{P}_0$. Here, we denote $\mathbb{P}_0'$ as the distribution after employing Laplace mechanism, and p_0' as its density of probability function accordingly. According to the definition of ϵ-DP in Sect. 2.1.1, we have $p_0' \leq p_0 e^{\epsilon}$. Thus, the data-driven based PSP's revenue maximization under ϵ-DP, i.e., 3DPP problem, can be reformulated as follows.

$$\max_{\gamma} \quad -\sum_{j=1}^{M} c W_j \gamma_j + \sum_{j=1}^{M} c W_j$$

$$+ \min_{\mathbb{P}\in\mathscr{D}'} b\mathbb{E}_{\mathbb{P}}\Big(\min(\sum_{j=1}^{M} W_j \gamma_j, \xi)\Big), \tag{3.5}$$

s.t.:   (3.2), (3.3)

$$\mathscr{D}' = \{\mathbb{P} : d_{\zeta}(\mathbb{P}_0', \mathbb{P}) \leq \theta\}, \tag{3.6}$$

$$d_{\zeta}(\mathbb{P}_0', \mathbb{P}) = \sup_{h\in\mathscr{H}} \left| \int_{\Omega} h d\mathbb{P}_0' - \int_{\Omega} h d\mathbb{P} \right|. \tag{3.7}$$

3.3 3DPP Proof and Solutions

This section is organized as follows. Since in Sect. 2.2, we present how to determine converge rate under ζ-structure probability structure. In this section, we show the relation between DP parameter ϵ and distribution tolerance θ in ζ-structure probability structure, and prove our DP mechanism satisfies the requirement of data-driven, which is $d_\zeta(\mathbb{P}_0', \mathbb{P}) \leq \theta$. Second, we reformulate the problem under ζ-structure probability metrics, and convert it to a traditional two-stage robust optimization. We develop algorithms to solve the problem w.r.t. different probability metrics.

3.3.1 Converge Rate Under ζ-Structure Probability Metrics with ε-DP

Then we prove the converge rate between distribution with Laplace mechanism $\mathbb{P}_0'$ and real distribution $\mathbb{P}$ under Kantorovich metric as follows.

Proposition 1 *For a general dimension case (i.e., n≥1),*

$$\mathbb{P}(d_K(\mathbb{P}_0', \mathbb{P}) \leq \theta) \geq 1 - \exp\left(-\frac{\theta^2 V}{2\varnothing^2} - \epsilon\right). \tag{3.8}$$

Proof Let us define a set

$$\mathscr{B} := \{\mu \in \mathscr{P}(\Omega) : d_k(\mu, \mathbb{P}) \geq \theta\}, \tag{3.9}$$

where $\mathscr{P}(\Omega)$ is the set of all probability measures defined on Ω. Let $\mathscr{C}(\Omega)$ be the set of bounded continuous function $\phi \to R$. Therefore, following the definitions, for each $\phi \in \mathscr{C}(\Omega)$, we have

$$\mathbb{P}\left(d_K(\mathbb{P}_0', \mathbb{P}) \geq \theta\right) = Pr(\mathbb{P}_0' \in \mathscr{B}) \tag{3.10}$$

$$\leq Pr\left(\int_\Omega \phi d\mathbb{P}_0' \geq \inf_{\mu \in \mathscr{B}} \int_\Omega \phi d\mu\right) \tag{3.11}$$

$$\leq \exp\left(-V \inf_{\mu \in \mathscr{B}} \int_\Omega \phi d\mu\right) E\left(e^{V \int_\Omega \phi d\mathbb{P}_0 e^\epsilon}\right) \tag{3.12}$$

$$= \exp\left(-V \inf_{\mu \in \mathscr{B}} \left\{\int_\Omega \phi d\mu - \frac{1}{V} \log E\left(e^{V \int_\Omega \phi d\mathbb{P}_0 e^\epsilon}\right)\right\}\right)$$

$$= \exp\left(-V \inf_{\mu \in \mathscr{B}} \left\{\int_\Omega \phi d\mu - \frac{1}{V} \log E\left(e^{\sum_{i=1}^V e^\epsilon \phi(\xi^i)}\right)\right\}\right) \tag{3.13}$$

$$= \exp\left(-V \inf_{\mu \in \mathscr{B}} \left\{\int_\Omega \phi d\mu - \log \int_\Omega e^\epsilon e^\phi d\mathbb{P}\right\}\right) \tag{3.14}$$

$$= \exp\left(-V \inf_{\mu \in \mathscr{B}} \left\{\int_\Omega \phi d\mu - \log \int_\Omega e^\phi d\mathbb{P} - \epsilon\right\}\right), \tag{3.15}$$

where (3.10) follows the definition of $\mathscr{B}$, inequality (3.11) is from the fact that $\mathbb{P}_0 \in \mathscr{B}$, and μ is the one distribution in $\mathscr{B}$ that satisfies the minimum of $\int_\Omega \phi d\mu$,

(3.12) follows from the Chebyshev's exponential inequality [4], and (3.13) follows from the definition of $\mathbb{P}_0$.

Now we define $\Delta(\mu) := \sup_{\phi \in \mathscr{C}(\Omega)} \int_\Omega \phi d\mu - \log \int_\Omega e^\phi d\mathbb{P}$. Thus, following the definition of $\mathscr{C}(\Omega)$, there exists a series ϕ_n such that $\lim_{n \to \infty} \int_\Omega \phi d\mu - \log \int_\Omega e^\phi d\mathbb{P} = \Delta(\mu)$. For any small positive number $\theta' > 0$, there exists a constant number n_0 such that $\Delta(\mu) - (\int_\Omega \phi_n d\mu - \log \int_\Omega e_n^\phi d\mathbb{P}) \leq \theta'$ for any $n \geq n_0$. Therefore, according to (3.15), we use substitute ϕ_n for ϕ, then we have

$$Pr(\mathbb{P}_0' \in \mathscr{B})$$

$$\leq \exp\left(-V \inf_{\mu \in \mathscr{B}} \left\{ \int_\Omega \phi d\mu - \log \int_\Omega e^\phi d\mathbb{P} - \epsilon \right\}\right) \tag{3.16}$$

$$\leq \exp\left(-V \inf_{\mu \in \mathscr{B}} \left\{ \Delta(\mu) - \epsilon - \theta' \right\}\right) \tag{3.17}$$

According to Lemma 6.2.13 in [8], we have

$$\Delta(\mu) = d_{KL}(\mu, \mathbb{P}) \tag{3.18}$$

where $d_{KL}(\mu, \mathbb{P})$ is the discrete case KL-divergence defined as $\sum_i ln(p_i/\mu_i)p_i$. For the case $\mu \in \mathscr{B}$, with (3.9), we have $d_K(\mu, \mathbb{P}) \geq \theta$. Moreover, in "Particular case 5" in [1], we have

$$d_K(\mu, \mathbb{P}) \leq \varnothing \sqrt{2 d_{KL}(\mu, \mathbb{P})} \tag{3.19}$$

hold for $\forall \mu \in \mathscr{P}(\Omega)$. Consequently, following (3.19), we have

$$d_{KL}(\mu, \mathbb{P}) \geq \theta^2 / \left(2\varnothing^2\right). \tag{3.20}$$

Combining (3.17), (3.18), (3.20), we have

$$Pr(\mathbb{P}_0' \in \mathscr{B}) \leq \exp\left(-V \left(\frac{\theta^2}{2\varnothing^2} - \epsilon - \theta'\right)\right). \tag{3.21}$$

Let $\theta' = \lambda/V$ for any arbitrary small positive λ. Then, we have

$$P\left(d_k(\mathbb{P}_0', \mathbb{P}) \geq \theta\right)$$

$$= Pr\left(\mathbb{P}_0' \in \mathscr{B}\right) \leq \exp\left(-V \left(\frac{\theta^2}{2\varnothing^2} - \epsilon\right) + \lambda\right). \tag{3.22}$$

Since λ can be arbitrarily small, we have $\mathbb{P}(d_k(\mathbb{P}_0', \mathbb{P} \leq \theta)) \geq 1 - \exp(-\frac{\theta^2}{2\varnothing^2}V + V\epsilon)$.

With convergence rate (3.22), we can calculate the tolerance θ accordingly. For instance, in Kantorovich metric, we assume the confidence level is η. Therefore $\mathbb{P}(d_u(\mathbb{P}_0, \mathbb{P} \leq \theta)) \geq 1 - \exp(-\frac{\theta^2}{2\varnothing^2}V + V\epsilon) = \eta$ according to (3.22), and $\theta = \varnothing\sqrt{2log(e^{\epsilon V}/(1-\eta))/V}$.

Similar proof is applicable for other metrics. For example, following the proof procedure of Proposition 1 in our work and using Corollary 1 in [9], it is easy to prove that under Fortet–Mourier metric, we have

$$\mathbb{P}(d_{FM}(\mathbb{P}'_0, \mathbb{P}) \leq \theta) \geq 1 - \exp\left(-\frac{\theta^2 V}{2\varnothing^2 \Lambda^2} + \epsilon V\right), \tag{3.23}$$

where $\Lambda = \max\{1, \varnothing^{p-1}\}$. Due to the page limits, we omit the detailed proof procedure.

3.3.2 Problem Reformulation Under ζ-Probability Metrics, and Solutions

We denote $x = \sum_{j=1}^{M} W_j \gamma_j$, $\alpha = \sum_{j=1}^{M} W_j$ where α is a constant. The sample space is $\Omega = \{\xi_1, \xi_2, \cdots, \xi_N\}$. Then the formulation can be simplified as

$$\max_{x} \quad -cx + \min_{p_i} b \sum_{i=1}^{N} p_i \left(\min(x, \xi_i)\right) + c\alpha \tag{3.24}$$

$$\text{s.t.} \quad x \in [0, \alpha], \tag{3.25}$$

$$\sum_{i} p_i = 1, \tag{3.26}$$

$$\max \sum_{i=1}^{N} h_i p'_{0_i} - \sum_{i=1}^{N} h_i p_i \leq \theta, \forall h_i : ||h||_\zeta \leq 1, \tag{3.27}$$

where the $|h||_\zeta$ is defined according to different metric. In Kantorovich metric, $|h_x - h_y| \leq \rho(\zeta^x, \zeta^y)$. The constraint (3.26), (3.27) can be summarized as $\sum_i a_{il} h_i \leq b_{il}, l = 1, \cdots, L$. To reformulate the constraint, we consider the problem

$$\min_{h_i} \quad \sum_{i=1}^{N} h_i p'_{0_i} - \sum_{i=1}^{N} h_i p_i, \tag{3.28}$$

$$\text{s.t.} \quad \sum_{i=1}^{N} a_{il} h_i \leq b_{il}, l = 1, \cdots, L. \tag{3.29}$$

Its dual problem is represented as

$$\min \quad \sum_{l=1}^{L} b_l u_l, \tag{3.30}$$

$$\text{s.t.} \quad \sum_{l=1}^{l} a_{il} u_l \geq p'_{0_i} - p_i, \forall i = 1, \cdots, N, \tag{3.31}$$

where u is the dual variable. Accordingly, the formulation can be reformulated as follows

$$\max_{x} \quad -cx + \min_{p_i} b \sum_{i=1}^{N} p_i \left(\min(x, \xi_i) \right) + c\alpha, \tag{3.32}$$

$$\text{(SP-M)} \quad \text{s.t.} \quad x \in [0, \alpha], \tag{3.33}$$

$$\sum_{i=1}^{N} p_i = 1, \sum_{l=1}^{l} b_l u_l \leq \theta, \tag{3.34}$$

$$\sum_{l=1}^{L} a_{il} u_l \geq p'_{0_i} - p_i, \forall i = 1, \cdots, N. \tag{3.35}$$

For the uniform metric, we can have the reformulation from the Uniform metric definition

$$\max_{x} \quad -cx + \min_{p_i} b \sum_{i=1}^{N} p_i \left(\min(x, \xi_i) \right) + c\alpha \tag{3.36}$$

$$\text{(SP-U)} \quad \text{s.t.} \quad x \in [0, \alpha], \tag{3.37}$$

$$\sum_{i=1}^{N} p_i = 1, \tag{3.38}$$

$$\left| \sum_{i=1}^{l} (p'_{0_i} - p_i) \right| \leq \theta, \forall l = 1, \cdots, L. \tag{3.39}$$

The formulation SP-M and SP-U can be solved by L-shape algorithm which is described in [10]. We summarize the procedure of solving the 3DPP problem in Algorithm 1.

Algorithm 1 Procedure of Solving $3DPP$

1: **Input:** Historical data $\xi_1, \xi_2, \cdots, \xi_N$ from sample SUs. Set ϵ as the privacy parameter. Set η as the confidence level of D.
2: **Out:** Objective value of the η.
3: STED receives the number of sampled SUs under different traffic demand, i.e., $\xi_1, \cdots, \xi_N$.
4: STED adds Laplace noise to the original data set of sample SUs. $\xi_n' = \xi_n + (Y_1, \cdots, Y_k)$, where Y_i are i.i.d random variables drawn from $\text{Lap}(\Delta f / \epsilon)$.
5: STED reports the processed data ξ_n' to PSP.
6: Obtain the reference distribution $\mathbb{P}_0'(\xi)$ and tolerance θ based on the data received from STED.

7: STED uses the reformulation (SP-M) or (SP-U) to solve the problem.
8: Output the solution.

3.4   Performance Evaluation

3.4.1   Simulation Setup

For illustrative purposes, we consider a spectrum trading market with 500 SUs. We assume the true traffic demand of all SUs follows a discrete distribution: 100M with probability 0.4 and 200M with probability 0.6, respectively. Total available spectrum resources aggregated by the PSP is 300M. In addition, we set the fixed wholesale price for the spectrum sold to the SSP to be $3/MHz, and the spot price for the spectrum sold directly to SUs to be $5/MHz.

3.4.2   Privacy and Performance Analysis

First, the confidence level η is set to be 90% and the size of sampled SUs varies from 10 to 120. We study the data-driven algorithm without DP. The results are shown in Fig. 3.2a. After collecting traffic demand of sample SUs, the STED does not add Laplace noises, and submits the true reference distribution directly to the PSP. From the results in Fig. 3.2a, it can be observed the total revenue of the PSP increases when the size of sample SUs increases, regardless of the distance metrics adopted. The intuition behind the result is that, as the size of sampled SUs, the value θ decreases, which stands for the distance between true distribution and reference distribution. As a result, the solutions are moving closer to the optimal one. It is also shown in Fig. 3.2a that the gap between total revenue under the Fortet–Mourier metric and the Kantorovich metric is very small, when the number of sampled SUs is over 100. When the number of sampled SU is 120, the results under all metrics are close to the optimal one. Besides, we study the 3DPP's performance in Fig. 3.2b. Compared with results in Fig. 3.2a, it can be observed that the total revenue of the PSP with 3DPP is less than that without ϵ-DP when the number of sampled SUs is small, but becomes close to each other, or even to the optimal revenue when the size of sampled SUs increases. That means it incurs some cost to involve

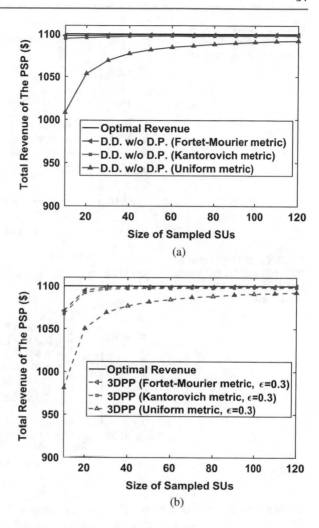

Fig. 3.2 Total revenue of PSP under different probability distance metrics. (**a**) Data-Driven spectrum trading without ϵ-DP. (**b**) 3DPP

ϵ-DP for the sampled SUs' traffic demands, especially when the number of sampled SUs is small. But this impact significantly diminishes when the number of samples increases. That also implies that the proposed 3DPP scheme can still successfully captures the characteristics of whole data set, i.e., the demand distribution of all SUs, while preserving individual sampled SU's traffic profile privacy. Moreover, from Fig. 3.5, we found the Fortet–Mourier metric is a more applicable metric, since the simulation results is more closer to the optimal revenue. In addition, we explore the impact of DP parameter ϵ in Figs. 3.3, 3.4 and 3.5. We choose four different ϵ values, i.e., 0.7, 0.5, 0.3, 0.2, respectively, and study its impact under different metrics. We find that as the ϵ decreases, the total revenue of PSP decreases under all metrics. The reason is, ϵ stands for the upper bound of privacy loss. It means, when ϵ is smaller, the mechanism yields better privacy, and less accurate responses which

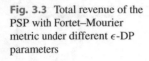

Fig. 3.3 Total revenue of the PSP with Fortet–Mourier metric under different ϵ-DP parameters

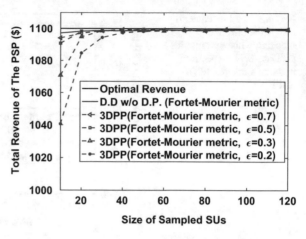

Fig. 3.4 Total revenue of the PSP with Kantorovich metric under different ϵ-DP parameters

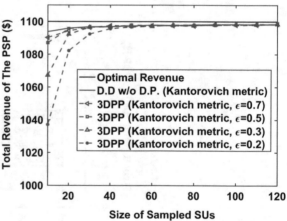

leads to less revenue of the PSP. It also can be observed that, when size of sampled SUs is less than 60, the gaps of total revenue under different ϵ is large. When size of sampled SUs increases, the influence of ϵ is less, and the total revenue under 3DPP with different ϵ converges to the optimal one.

Last but not the least, we study the effect of confidence level on the 3DPP in Fig. 3.6. We set the number of sampled SUs as 40, and test four different confidence levels, i.e., 0.6, 0.7, 0.8, 0.9, respectively. From Fig. 3.6 we can observe that, as the confidence level increases, the gaps between the PSP's revenue of 3DPP and optimal one increases under all three metrics. The reason is that, as the confidence level η increases, the distance θ between reference distribution with ϵ-DP $\mathbb{P}'_0$ and true distribution $\mathbb{P}$ increases, and the true probability distribution of SUs traffic demands is more likely to be in the confidence set $\mathcal{D}$. That implies that distribution in set $\mathcal{D}$ which is not that close to $\mathbb{P}$ might be used to yield solutions. Therefore, the PSP's revenue performance degrades when confidence level increases.

Fig. 3.5 Total revenue of the PSP with uniform metric under different ϵ-DP parameters

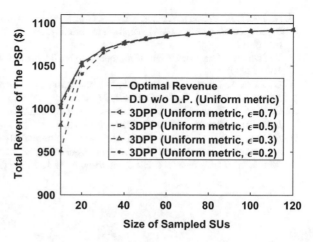

Fig. 3.6 Total revenue of the PSP with 3DPP under different confidence levels

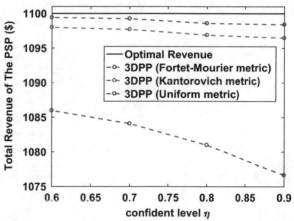

References

1. F. Bolley, C. Villani, Weighted csiszar-kullback-pinsker inequalities and applications to transportation inequalities. Ann. Fac. Sci. Toulouse **14**, 331–352 (2005)
2. L. Duan, J. Huang, B. Shou, Cognitive mobile virtual network operator: investment and pricing with supply uncertainty, in *Proceedings of IEEE Conference on Computer Communications, INFOCOM 2010*, San Diego, CA (2010)
3. C. Dwork, Differential privacy: a survey of results, in *International Conference on Theory and Applications of Models of Computation* (Springer, New York, 2008), pp. 1–19
4. J.M. Hammersley, D.C. Handscomb, *Monte Carlo Methods* (Methuen, London, 1964)
5. X. Jin, Y. Zhang, Privacy-preserving crowdsourced spectrum sensing, in *Proceeding of the IEEE International Conference on Computer Communications (INFOCOM)* (2016), pp. 1–9
6. X. Li, H. Ding, M. Pan, Y. Sun, Y. Fang, Users first: service-oriented spectrum auction with a two-tier framework support. IEEE J. Sel. Areas Commun. Spectr. Shar. Aggregation Future Wirel. Netw. **34**(11), 2999–3013 (2016)

7. M. Pan, P. Li, Y. Song, Y. Fang, P. Lin, S. Glisic, When spectrum meets clouds: optimal session based spectrum trading under spectrum uncertainty. IEEE J. Sel. Areas Commun. **32**(3), 615–627 (2014)
8. U. Schmock, Large deviations techniques and applications. J. Am. Stat. Assoc. **95**(452), 1380–1381 (2000)
9. C. Zhao, Y. Guan, Data-driven risk-averse two-stage stochastic program with ζ-structure probability metrics. Optim. Online **2**, 9 (2015)
10. C. Zhao, Y. Yuan, Data-driven stochastic unit commitment for integrating wind generation. IEEE Trans. Power Syst. **31**(4), 2587–2596 (2016)
11. R. Zhu, Z. Li, F. Wu, K. Shin, G. Chen, Differentially private spectrum auction with approximate revenue maximization, in *Proceedings of ACM International Symposium on Mobile Ad Hoc Networking and Computing, ACM MobiHoc* (2014), pp. 185–194

Optimization for Utility Providers with Privacy Preservation of Users' Energy Profile

4

Abstract

Smart meters migrate conventional electricity grid into digitally enabled SG, which is more reliable and efficient. Fine-grained energy consumption data collected by smart meters helps utility providers accurately predict users' demands and significantly reduce power generation cost, while it imposes severe privacy risks on consumers and may discourage them from using those "espionage meters". To enjoy the benefits of smart meter measured data without compromising the users' privacy, in this chapter, we try to integrate DDP techniques into data-driven optimization, and propose a novel scheme that not only minimizes the cost for utility providers but also preserves the DDP of users' energy profiles. Briefly, we add differential private noises to the users' energy consumption data before the smart meters send it to the utility provider. Due to the uncertainty of the users' demand distribution, the utility provider aggregates a given set of historical users' differentially private data, estimates the users' demands, and formulates the data-driven cost minimization based on the collected noisy data. We also develop algorithms for feasible solutions, and verify the effectiveness of the proposed scheme through simulations using the simulated energy consumption data generated from the utility company's real data analysis.

4.1 Network Model

In smart grid, due to the two-way communication of electricity and information, the utility providers are supposed to profile users' demand in order to efficiently balance the supply and demand, at the meanwhile, reduce the cost [3–5]. In practice, there exists a significant amount of historical data about consumers' demand. With a given set of energy profiles collected by smart meters, at utility provider side,

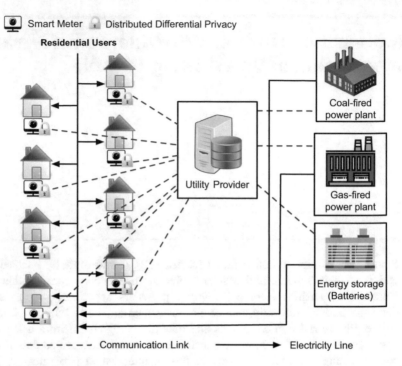

Fig. 4.1 Network overview

data-driven is applied to forecast the demand. In addition, we assume the utility provider is not trustworthy and the users add the differential noises by themselves [1, 7]. In our work, we make the user perform DDP algorithm, which is introduced in detail in Sect. 2.1.2 to randomize the true data and send the modified value to the utility provider. As shown in Fig. 4.1, with the smart meters, the residential users first process their true energy demand with the DDP algorithm [6]. The utility provider collects the data from the given set of users and applies data-driven model to estimate the energy demands and determine the amount of supply.

In our architecture, the utility provider predicts the future demand from history data of a given set of customers' noisy demand, construct a reference distribution and predict the total energy demand for all customers. Since the reference distribution cannot present 100% unique true demand distribution, the scheduled energy supply may not meet the demand of all customers. Under this scenario, when the supply and demand are not matched, the quick-response efficient gas-fired power plant or pumped-storage electrical power station will be started up. We assume the given set of residential users is $\mathcal{N} = \{1, \cdots, j, \cdots, N\}$ and the real demand for each user is U_j. There are several backup gas-fired power plants/energy storage from a set $\mathcal{M} = \{1, \cdots, i, \cdots, M\}$ controlled and operated by the utility provider. Each has the capacity of c_i unit of electricity power provided to the consumers.

4.1.1 Data-Driven Prediction

The traditional two-stage stochastic programming approach under our scenario assumes the distribution of the consumers' energy demand is known. However, in reality, the distribution for the forecasting energy demand is actually uncertain. Instead, only a series of historic consumer's energy profile data are available. In this chapter, we employ a data-driven approach, i.e., the risk-averse stochastic optimization approach (RA-SP) allowing distribution ambiguity, to characterize the uncertainty of forecasting energy demand. In the proposed method, we build the reference distribution from a given set of empirical consumers' data. Since the reference distribution from empirical data might be different from the true distribution, we employ statistical inference and define confidence sets $\mathscr{D}$ corresponding to a given tolerance θ. It allows the distribution ambiguity to be within confident sets $\mathscr{D}$ with a certain confidence level (e.g., 99%). We employ two norms, L_1 and L_∞ norms, to construct two types of confidence sets D_1 and D_∞. As the number of sampled days is sufficiently large, the reference distribution converges to the true distribution under both two norms. In the following, we describe the two confident sets D_1 and D_∞ as:

$$\mathscr{D}_1 = \left\{ P \in \mathbb{R}_+^K \,\middle|\, \|P - P_0\|_1 \le \theta \right\}$$

$$= \left\{ P \in \mathbb{R}_+^K \,\middle|\, \sum_{k=1}^{K} \left| p_k - p_k^0 \right| \le \theta \right\} \tag{4.1}$$

and

$$\mathscr{D}_\infty = \left\{ P \in \mathbb{R}_+^K \,\middle|\, \|P - P_0\|_\infty \le \theta \right\}$$

$$= \left\{ P \in \mathbb{R}_+^K \,\middle|\, \max_{1 \le k \le K} |p_k - p_k^0| \le \theta \right\}. \tag{4.2}$$

Under these two norms, the formulated problem can be obtained as a mixed integer linear programming eventually.

For a given set of processed energy profiles data (assuming there are historical data samples), it is easy for us to construct a histogram to fit all the energy profiles data. For example, we can set K intervals to fit the predicted total energy demand of sampled days in each interval to be $L_1, L_2, \cdots$, and L_K with $L = \sum_{k=1}^{K} L_k$. Based on this, we can construct an reference distribution for the uncertain total energy demand of all consumers in particular time period of a day as $p_1^0 = L_1/L$, $p_2^0 = L_2/L, \cdots$, and $p_K^0 = L_K/L$. For simplicity, we let $P_0 = p_1^0, p_2^0, \cdots, p_K^0$ represent the corresponding reference distribution.

The two distribution sets under $Norm_1$ and $Norm_\infty$ are built based on a given confidence level and the amount of available historical data. For instance, β is set to represent the confidence level and $\beta = 98\%$ indicates that the ambiguous distribution P has at least 98% chance in the given set. In (4.1) and (4.2), θ denotes the tolerance value, which is derived from the confident set β and the number of historical data. Intuitively, the more historical data we have, the more "closer" between the reference distribution and true distribution. From [8], we can explore the precise relationship between the tolerance θ and the number of historical data L. The propositions are shown as follows:

Proposition 1 *Supposing there are L number of historical samples, and K intervals, the convergence rate between P and P_0 under L_1 norm is:*

$$Pr\{P \in \mathbb{R}_+^K | \|P - P_0\|_1 \leq \theta\} \geq 1 - 2K \exp(-2L\theta/K).$$

Proposition 2 *Supposing there are L number of historical samples, and K intervals, the convergence rate between P and P_0 under L_∞ norm is:*

$$Pr\{P \in \mathbb{R}_+^K | \|P - P_0\|_\infty \leq \theta\} \geq 1 - 2K \exp(-2L\theta).$$

We can derive the relation between confidence level β and the tolerance θ from above as follows:

$$\theta \text{ for } L_1 \text{ norm} : \theta_1 = \frac{K}{2L} \log \frac{2K}{1 - \beta}, \tag{4.3}$$

$$\theta \text{ for } L_\infty \text{ norm} : \theta_\infty = \frac{1}{2L} \log \frac{2K}{1 - \beta}. \tag{4.4}$$

From (4.3) and (4.4), it is easy to observe that, as the size of historical data L increases to ∞, both tolerance θ_1 and θ_∞ decrease to 0. Therefore, the confidence sets $\mathscr{D}_1$ and $\mathscr{D}_\infty$ become singleton, and the corresponding risk-averse two stage stochastic problem becomes the traditional two-stage stochastic problem.

4.1.2   Cost Minimization Problem Formulation

The collected data with DDP from smart meters is aggregated by the utility provider. Because the distribution of the customers' demand is uncertain, in order to efficiently balance the supply and demand, the utility provider employs data-driven approach to forecasting the customers' future demand. As the reference distribution constructed from the collected data cannot present the unique true distribution of the customers' demand. In order to match the supply and demand, the quick-response gas-fired power plants or pumped-storage electrical power stations are used. At the same time, on the utility provider side, the cost is supposed to be minimized. Consequently, the cost minimization problem for utility provider can be formulated as follows:

$$\min_{x,y} \sum_i^N F_i y_i + \mathbb{E}_\mathbb{P}\left[\sum_i^M T_i x_i(\xi) \right], \tag{4.5}$$

s.t.:

$$\sum_i^M x_i(\xi) \le c_i y_i \quad \forall i, \tag{4.6}$$

$$\sum_i^M x_i(\xi) = \sum_j U_j - d(\xi), \tag{4.7}$$

$$x_i(\xi) \ge 0, \ y_i \in \{0, 1\} \quad \forall i, j, \tag{4.8}$$

In the formulation, (4.6) indicates the energy generated from gas-fired power plant i should not exceed its capacity and (4.7) indicates the total number of energy generated from all gas-fired power plants is the gap between overall real demand and uncertain predicted demand from energy utility. The opening price for each gas-fired power plant is represented by F_i, y_i is a binary variable indicating if gas-fired power plant i is open, T_i denotes the purchase price of each unit from power plant i, x_i is the energy generated from the power plant i, c_i expresses the capacity of each power plant and U_j means real demand from each consumer j in the particular time period.

Since we add noise in the processed energy profile, the distribution of real demand is ambiguous. Therefore, we construct the confident set $\mathcal{D}$, and let $P \in \mathcal{D}$ so as to minimize the total cost under the worst-case distribution realization in $\mathcal{D}$. The detailed formulation is described as follows:

$$\min_y \sum_i^M F_i y_i + \max_{p_k} \sum_k^K p_k \min_x \sum_i^M T_i x_i(\xi_k), \tag{4.9}$$

s.t.: (4.6)–(4.8),

$$\sum_{k=1}^K p_k = 1, \tag{4.10}$$

$$P \in \mathcal{D}. \tag{4.11}$$

4.1.3 Solution to the Optimization Problem

The Benders' decomposition algorithm [2] is exploited to solve the problem into global optimality. Since for each scenario ξ_k, the second-stage optimization problem $\min_x \sum_i^M T_i x_i(\xi^k)$ of (4.9) is independent of ξ_i for $i \neq k$. Consequently, the minimization operation can be put before the summation, i.e, the objective function (4.9) can be written as

$$\min_y \sum_i^M F_i y_i + \max_{p_k} \min_x \sum_{k=1}^K p_k \sum_i^M T_i x_i(\xi_k), \qquad (4.12)$$

s.t.:   (4.6)–(4.8), (4.10), (4.11).

We can calculate the second-stage minimization problem by solving its dual. The dual subproblem and dual variables λ, v associated with constraints are given by

$$\max_{\lambda, v} \sum_{k=1}^K \left[\lambda_k (U - d(\xi_k)) - \sum_i^M c_i y_i v_k^i \right] \qquad (4.13)$$

s.t.: $\lambda_k - v_k^i \leq p_k T_i, \forall i, k$ \qquad (4.14)

$v_k^i \geq 0, \forall i, k.$ \qquad (4.15)

The dual variables corresponding to scenario k for constraints (4.6)–(4.8) are v_k^i and λ_k, respectively. Because of the duality property, the optimal objective of the primal problem is equivalent to the dual problem. It is obvious that the maximization operation in the primal formulation can be combined with the dual second-stage problem. The second-stage max-min problem can be obtained as

$$\psi(y) = \max_{p_k} \min_x \sum_{k=1}^K p_k \sum_i^M T_i x_i(\xi_k)$$

$$= \max_{p_k, \lambda, v} \sum_{k=1}^K \left[\lambda_k (U - d(\xi_k)) - \sum_i^M c_i y_i v_k^i \right], \qquad (4.16)$$

s.t.: $\lambda_k - v_k^i \leq p_k T_i, \forall i, k,$ \qquad (4.17)

$v_k^i \geq 0, \forall i, k,$ \qquad (4.18)

$$\sum_{k=1}^K p_k = 1, \qquad P \in D. \qquad (4.19)$$

Under the L_∞ norm case, the constraint (4.19) represents

$$\max_{1 \leq k \leq K} |p_k - p_k^0| \leq \theta, \tag{4.20}$$

which is equal to

$$|p_k - p_k^0| \leq \theta, \forall k. \tag{4.21}$$

Under the L_1 norm case, the constraint (4.19) represents

$$\sum_{k=1}^{K} |p_k - p_k^0| \leq \theta. \tag{4.22}$$

We denote α as the second-stage worst case energy cost. Then by applying feasibility cut and optimality cut iteratively, we can solve the master problem which is reformulated as follows.

$$\min_{y \in \{0,1\}} \sum_{i}^{N} F_i y_i + \alpha$$

$$\text{s.t.: Feasibility cuts,}$$

$$\text{Optimality cuts.}$$

- *Feasibility Cuts*: We use the L-shaped method to generate feasibility cuts. We formulate the feasibility check problem as follows to check constraints (5.2) and (5.3):

$$\min_{\gamma,x} \sum_{k=1}^{K} \left(\sum_{i=1}^{M} \gamma_k^{1i} + \gamma_k^2 + \gamma_k^3 \right) \tag{4.23}$$

$$\text{s.t.} \quad \gamma_k^{1i} - X_i(\xi_k) \geq -c_i y_i, \forall i, k, \tag{4.24}$$

$$\gamma_k^2 - \sum_{i=1}^{M} X_i(\xi_k) \geq -(U - d(\xi_k)), \forall k, \tag{4.25}$$

$$\gamma_k^3 - \sum_{i=1}^{M} X_i(\xi_k) \geq (U - d(\xi_k)), \forall k, \tag{4.26}$$

$$X_i(\xi_k) \geq 0, \gamma_k^{1i}, \gamma_k^2, \gamma_k^3 \geq 0, \forall i, k. \tag{4.27}$$

Its dual problem can be obtained as follows:

$$\omega(y) = \tag{4.28}$$

$$\max_{\hat{\lambda}, \hat{\mu}, \hat{v}} \sum_{k=1}^{K} \left[-\hat{\lambda}_k \left(U - d(\xi_k) \right) + \hat{\mu}_k \left(U - d(\xi_k) \right) - \sum_{i=1}^{M} \hat{v}_k^i c_i y_i \right] \tag{4.29}$$

$$\text{s.t.} \quad -\hat{\lambda}_k + \hat{\mu}_k + \hat{v}_k^i \le 1, \forall i, k, \tag{4.30}$$

$$\hat{\lambda}_k, \hat{\mu}_k, \hat{v}_k^i \in [0, 1], \forall i, k, \tag{4.31}$$

where dual variables $\hat{v}_k^i$, $\hat{\lambda}_k$ and $\hat{\mu}_k$ correspond to the kth scenario for constraints (4.24), (4.25) and (4.26), respectively. Therefore, the feasibility check is performed as follow steps:

1. If $\omega(y) = 0$, the first stage solution is feasible.
2. If $\omega(y) \ge 0$, a feasible cut is generated in the following form:

$$\sum_{k=1}^{K} \left[-\hat{\lambda}_k \left(U - d(\xi_k) \right) + \hat{\mu}_k \left(U - d(\xi_k) \right) - \sum_{i=1}^{M} \hat{v}_k^i c_i y_i \right] \le 0. \tag{4.32}$$

- *Optimality Cuts*: At each iteration, we get y and α after solving the master problem. Then we substitute y into the subproblem and obtain $\psi(y)$. If $\psi(y) \le \alpha$, we claim we find the optimal solution. If not, which means $\psi(y) > \alpha$, we generate optimal cut in the following form and add it into the master problem:

$$\sum_{k=1}^{K} \left[\lambda_k \left(U - d(\xi_k) \right) - \sum_{i} c_i y_i v_k^i \right] \le \alpha. \tag{4.33}$$

Finally, the optimality cuts and feasibility cuts ensure the Benders' decomposition algorithm converges to global optimality.

4.2 Performance Evaluation

In this section, we evaluate our proposed model and associated algorithms. The evaluation is accomplished in a computer equipped with Intel Core i7 CPU of 2.7 GHz. Due to the nondisclosure agreement, all results are computed from the simulated data that are generated according to the real data analysis. The proposed algorithms can be directly applied to real data without modification. The utility provider processes 10,000 consumers energy cost per hour (from 8 pm to 9 pm) for 300 days. The consumers implement DDP algorithm to their energy profiles. We assume the total consumers' energy consumption is 6×10^4. In our model, there are three gas-fired power plants, each capacity is 4×10^4; 5×10^4 and

1×10^4, accordingly. In addition, the open cost for each gas-fired power plant is 1, and the unit wholesale price for energy is 1 per unit.

We set the confidence level β to 80% and study the data-driven algorithm without DDP. The results are shown in Figs. 4.2, 4.3, and 4.4. From Fig. 4.2, we can observe that the distribution of consumers' energy demand is very close after integrating distributed differential privacy. We notice that as ϵ get smaller, the privacy is higher,

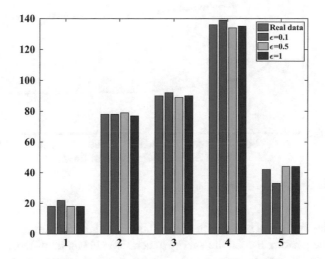

Fig. 4.2 Distribution of consumers' energy demand under different ϵ

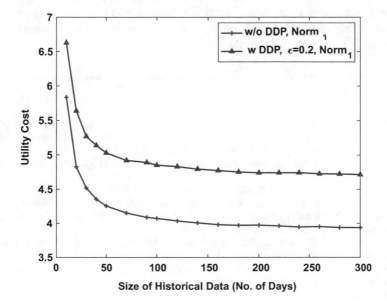

Fig. 4.3 Total cost under L_1 norm

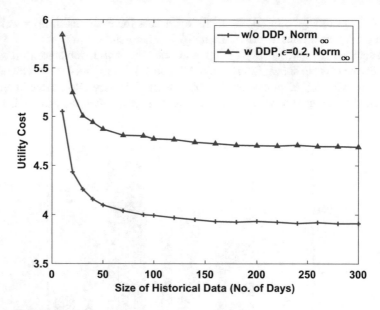

Fig. 4.4 Total cost under L_∞ norm

therefore, the difference between the distributions is higher too. In Figs. 4.3 and 4.4, we obtain the performance of the energy generation cost under different forms. It is shown that under both L_1 norm and L_∞ norm, the cost is lower as we have more historical data. It means that with more historical data processed by utility provider, the distribution is more accurate. It can also be observed that the utility cost under DDP algorithm is worse than the performance without preservation privacy, which represents the trade-off between the privacy and utility.

References

1. M.R. Asghar, G. Russello, B. Crispo, M. Ion, Supporting complex queries and access policies for multi-user encrypted databases, in *Proceedings of the ACM Cloud Computing Security Workshop, Co-located with CCS*, Berlin, Germany, November 2013
2. A.M. Geoffrion, Generalized benders decomposition. J. Optim. Theory Appl. **10**(4), 237–260 (1972)
3. IEEE, IEEE Spectrum (2018). Retrieved from https://spectrum.ieee.org
4. X. Lou, R. Tan, D.K. Yau, P. Cheng, Cost of differential privacy in demand reporting for smart grid economic dispatch, in *Proceeding of the IEEE International Conference on Communications (ICC'17)*, Paris, France, May 2017
5. Illinois Commerce Commission, NextGrid: Illinois utility of the future study (2018). Retrieved from https://nextgrid.illinois.gov
6. E. Shi, H. Chan, E. Rieffel, R. Chow, D. Song, Privacy-preserving aggregation of time-series data, in *Annual Network & Distributed System Security Symposium (NDSS)*, San Diego, CA, February 2011

7. O. Vukovic, G. Dan, R.B. Bobba, Confidentiality-preserving obfuscation for cloud-based power system contingency analysis, in *IEEE International Conference on Smart Grid Communications (SmartGridComm)*, Vancouver, Canada, December 2013
8. C. Zhao, Y. Guan, Data-driven stochastic unit commitment for integrating wind generation. IEEE Trans. Power Syst. **31**(4), 2587–2596 (2016)

Caching with Users' Differential Privacy Preservation in Information-Centric Networks

5

Abstract

Information-centric networking (ICN) is developed for the future Internet because of the tremendous increase of content demands in the Internet. In the ICN architecture, in-network storage for caching plays an important role in improving content delivery efficiency, scalability and availability. To enjoy the benefits of caching users' preferable contents without disclosing the users' privacy, in this chapter, we aim to integrate local differential privacy (LDP) techniques into data-driven optimization, and propose a novel scheme to allow content provider (CP) to collect the locally differentially private content preferences of a selected group of users, exploit data-driven approach to predict the content popularity, and offer the cache-enabled access points (APs) economic incentives to cache the selected preferable content. Here, optimized local hashing (OLH) is employed to locally add differential private noise to the users' preference content information and the noisy data is sent to the CP. Besides, we leverage data-driven methodology to predict the content popularity according to the constructed reference distribution of the given noisy preference content data from users. We formulate a data-driven caching revenue optimization, provide feasible solutions, and conduct simulations to show the effectiveness of the proposed scheme.

5.1 Network Model and Preliminaries

5.1.1 System Description

In our work, as shown in Fig. 5.1, we assume the content provider (CP), in the information-centric network (ICN) [1, 2, 5, 6], collects users' content preferences information with local differential noise, forecasts the content popularity by data-

© The Author(s), under exclusive license to Springer Nature Switzerland AG 2019   47
M. Pan et al., *Big Data Privacy Preservation for Cyber-Physical Systems*,
SpringerBriefs in Electrical and Computer Engineering,
https://doi.org/10.1007/978-3-030-13370-2_5

Fig. 5.1 System description

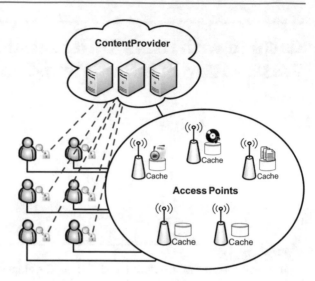

driven methodology, leases several storage for caching of the access points (APs) and offloads the popular contents in advance into the cache [1, 4, 5, 7]. Therefore, the heavy back haul load and congestion problem can be reduced. Additionally, the users apply the local differential privacy (LDP) protocols to add noise individually on their content preferences and send the modified value to the CP.

In our scheme, we assume the set of users is $\mathscr{U} = \{1, \cdots, u, \cdots, U\}$, the file is represented as f with size s_f and the real content preference of each user is r_u that is in the domain with size of F. There are several cache-enabled APs from a set $\mathscr{M} = \{1, \cdots, m, \cdots, M\}$ cooperated with the CP to provide high QoS. Each cache of the AP has the capacity of c_m unit and the price to lease each cache is k_m. With the LDP protocol, the users add noise locally to their content preference r_u, which is shown in Sect. 2.1.3 in detail. The CP constructs the reference content popularity probability $\mathbb{P}_0$ based on noisy content preference results and predicts the true popularity by data-driven approach. Hence, the storage for caching in APs is selected to lease by CP. Because of the uncertainty of reference distribution, the revenue maximization problem is formulated to determine the set of cache to be leased, which is illustrated in Sect. 5.1.3. Moreover, the Benders' decomposition is deployed to solve the proposed maximization problem.

5.1.2   Data-Driven Analysis of Content Popularity

Most works in the ICN assume that the distribution of content popularity is known as Zipf distribution. However, practically, it characterizes the statistical features in various geographical locations. Moreover, only historical data or real content preferences of users can be obtained by the CP to construct the reference distribution of content popularity. Therefore, in our work, we employ data-driven risk-averse

---

Algorithm 1 Algorithm for Obfuscation Strategy

1: **Input:** Survey data of user preference i.i.d drawn from the true distribution. The confident
   level of D is η.
2: **Output:** Objective value of the problem (5.1).
3: Obtain the reference distribution $\mathbb{P}_0$ and tolerance θ based on the historical data.
4: **if** The reference distribution and true distribution are under Kantorovich metric or Fortet-
   Mourier metric **then**
5:     Reformulate the problem to (5.13)–(5.15)
6:     Feasibility check master problem of (5.13)
7:     **if** Infeasible **then**
8:        Generate feasible cut for master problem
9:        go to line 6
10:    **end if**
11:    Feasibility check the subproblem of (5.13)
12:    **if** Infeasible **then**
13:       Generate optimal cut for subproblem
14:       go to line 6
15:    **end if**
16:    Stop and output solution
17: **else**
18:    Reformulate the problem to (5.16)–(5.18) under Uniform metric
19:    Solve the problem under bender decomposition algorithm, same as line 6 to line 16.
20:    Output the solution.
21: **end if**

---

stochastic optimization approach (RA-SP) to making a decision to lease cache-enabled APs under the uncertainty of predicting the content popularity. Specifically, a predefined distance measure $d(\mathbb{P}_0, \mathbb{P})$ is constructed on confident set D, where $\mathbb{P}_0$ is the reference distribution estimated from historical data, and $\mathbb{P}$ is the ambiguous distribution of users' content preferences distribution. The detail of data-driven analysis is illustrated in Sect. 2.2.

5.1.3 Caching Revenue Maximization Problem with Local Privacy Preservation

As we describe before, the CP collects user's noisy content preferences with LDP and aggregates the frequency estimation of each content [8]. Consequently, the CP is able to get the noisy content popularity represented as $r_u(f)$. In our work, according to the noisy content popularity $r_u(f)$, we assume there are F popular files in the set $\mathscr{F} = \{1, \cdots, f, \cdots, F\}$ selected to store in the cache-enabled APs, each of which has the capacity c_m. The binary parameter y_m is used to represent if an AP is leased by the CP with the price k_m. We denote the size of a file stored in a cache as s_{fm}. We represent the profit per unit size of backhaul load reduction as α. Hence, the total expected revenue of backhaul load reduction is $\alpha \sum_f^F \sum_m^M s_{fm}$. We randomly sample select a group of users of the CP and get the total download size of files. Since the uncertainty of the distribution of the content popularity, the total download size distribution can be denoted as $d(\xi)$. In addition, the profit of serving users per unit size of file is ϕ. In order to maximize revenue, the CP employs data-driven method as described in Sect. 2.2 to predict the real content popularity and selects cache-enabled APs from a given group [9]. Because of contribution of the APs, the backhaul load is reduced. Therefore, the revenue maximization problem for CP can be formulated as follows:

$$\max_{y} \sum_{m}^{M} -k_m y_m + \alpha \sum_{f}^{F} \sum_{m}^{M} s_{fm} + \mathbb{E}_{\mathbb{P}}[\phi d(\xi)], \tag{5.1}$$

$$\text{s.t.:} \quad \sum_{f}^{F} s_{fm} \leq c_m y_m, \forall m \tag{5.2}$$

$$y_m \in \{0, 1\}, \forall m, \tag{5.3}$$

In the formulation, (5.2) indicates the files store in one cache m should not exceed the capacity c_m of the cache and (5.3) indicates whether the cache m is leased by the CP. Since we add noise in the processed energy profile, the distribution of real demand is ambiguous. Therefore, we construct the confident set $\mathscr{D}$, and let $\mathbb{P} \in \mathscr{D}$ so as to minimize the total cost under the worst-case distribution realization in $\mathscr{D}$. The detailed formulation is described as follows:

$$\max_{y} \sum_{m}^{M} -k_m y_m + \alpha \sum_{f}^{F} \sum_{m}^{M} s_{fm} + \min_{\mathbb{P} \in \mathscr{D}} \mathbb{E}_{\mathbb{P}}[\phi d(\xi)], \tag{5.4}$$

$$\text{s.t.} \quad (5.2), (5.3),$$

$$\mathscr{D} = \{\mathbb{P} : d_{\zeta}(\mathbb{P}_0, \mathbb{P}) \leq \theta\}. \tag{5.5}$$

5.1.4 Solution to Caching Optimization Under Distribution Uncertainty

From Sect. 5.1.2, we explore how to solve (5.4). We assume the sample space is $\Omega = \{\xi^1, \xi^2, \cdots, \xi^N\}$. The formulation can be simplified as:

$$\max_{y} \sum_{m}^{M} -k_m y_m + \alpha \sum_{f}^{F} \sum_{m}^{M} s_{fm} + \min_{p_i} \sum_{i=1}^{N} p_i \left(\phi d(\xi^i)\right), \tag{5.6}$$

$$\text{s.t.} \quad (5.2), (5.3),$$

$$\sum_{n=1}^{N} p_i = 1, \tag{5.7}$$

$$\max_{h_i} \sum_{i=1}^{N} h_i p_i^0 - \sum_{i=1}^{N} h_i p_i \le \theta, \forall h_i : ||h||_\varsigma \le 1, \tag{5.8}$$

where $|h||_\varsigma$ is defined according to different metrics. For the Kantorovich metric, $|h_i - h_j| \le \rho(\xi^i, \xi^j)$. For the Fortet-Mourier metric, $|h_i - h_j| \le \rho(i, j) \max\{1, \rho(\xi^i, a)^{p-1}, \rho(\xi^j, a)^{p-1}\}$. The constraints (5.7)–(5.8) can be summarized as $\sum_{i=1}^{N} a_{ij} h_i \le b_j, j = 1, \cdots, J$. To reformulate the constraints, we consider the following problem:

$$\min_{h_i} \quad \sum_{i=1}^{N} h_i p_i^0 - \sum_{i=1}^{N} h_i p_i, \tag{5.9}$$

$$\text{s.t.} \quad \sum_{i=1}^{N} a_{ij} h_i \le b_j, j = 1, \cdots, J. \tag{5.10}$$

The dual problem can be formulated as:

$$\min_{u} \quad \sum_{j=1}^{J} b_j u_j, \tag{5.11}$$

$$\text{s.t.} \quad \sum_{j=1}^{J} a_{ij} u_j \ge p_i^0 - p_i, \forall i = 1, \cdots, N, \tag{5.12}$$

where u is the dual variable. Accordingly, the problem can be reformulated as follows under Kantorovich metric and Fortet-Mourier metric:

$$\max_{y} \sum_{m}^{M} -k_m y_m + \alpha \sum_{f}^{F} \sum_{m}^{M} s_{fm} + \min_{p_i} \sum_{i=1}^{N} p_i (\phi d(\xi^i)), \tag{5.13}$$

$$\text{s.t.} \quad (5.2), (5.3),$$

$$\sum_{n=1}^{N} p_i = 1, \sum_{j=1}^{J} b_j u_j \le \theta, \tag{5.14}$$

$$\sum_{j=1}^{J} a_{ij} u_j \ge p_i^0 - p_i, \forall i = 1, \cdots, N, \tag{5.15}$$

For the Uniform metric, we can have the reformulation from the Uniform metric definition:

$$\max_{y} \sum_{m}^{M} -k_m y_m + \alpha \sum_{f}^{F} \sum_{m}^{M} s_{fm} + \min_{p_i} \sum_{i=1}^{N} p_i (\phi d(\xi^i)), \tag{5.16}$$

s.t.   (5.2), (5.3),

$$\sum_{i=1}^{N} p_i = 1,,$$ (5.17)

$$\left| \left(p_i^0 - p_i \right) \right| \leq \theta, \forall i,$$ (5.18)

After reformulating the problem, we can solve the formulation (5.13)–(5.15) and (5.16)–(5.18) through Benders' decomposition algorithm [3]. The detailed algorithm is shown in Algorithm 1.

5.2   Performance Evaluation

In our simulation, we assume the CP provides service to 10,000 users and take a survey on the content preference from the selected users. The users implement local differential privacy protocol, then send the noisy preference result to the CP. To be specific, the sample users choose interested files from ten candidates, add noise and send back to the CP. The CP processes the results and estimates the maximum revenue. We assume there are three cache-enabled APs in our system, each capacity is 3 units, 4 units and 5 units, accordingly. To simplify simulation, each file is 4 units in our network.

We set the confidence level β to 99% and study the performance with and without integrating local differential privacy. In Fig. 5.2, we obtain the performance of the expected revenue under different metrics. It is shown that no matter under which metric, the expected revenue is higher and closer to optimal revenue as we have more data. It means that with more data processed by the CP, the distribution is more accurate. Figures 5.3, 5.4, and 5.5 shows the comparison under different local differential privacy levels. We can observe that the expected revenue after adding noise is worse than the performance without preserving privacy, which represents the trade-off between the privacy and utility. Moreover, it is observed that when ϵ increases from 0.5 to 1, the expected revenue increases under the same size of historical data, and closer to the performance without adding noise. The reason is that ϵ presents privacy level in differential privacy. When ϵ is smaller, it means the privacy level is higher, and the users would add more noise in the submitted data.

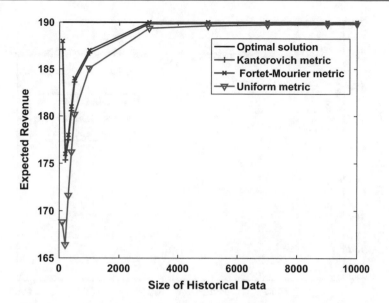

Fig. 5.2 Expected revenue without users' privacy preservation

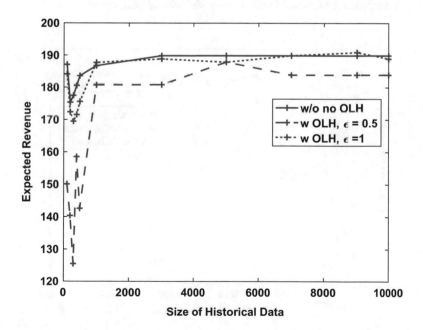

Fig. 5.3 Performance of Kantorovich metric under different privacy level

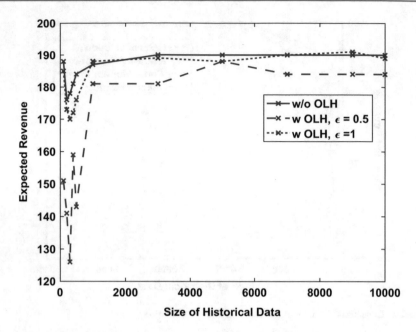

Fig. 5.4 Performance of Fortet-Mourier metric under different privacy level

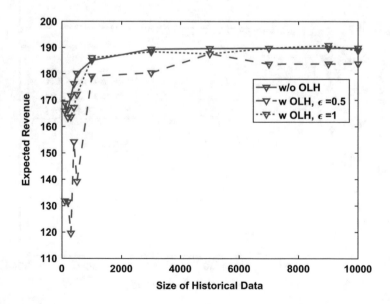

Fig. 5.5 Performance of Uniform metric under different privacy level

References

1. L.A. Adamic, B.A. Huberman, Zipf's law and the internet. Glottometrics **3**(1), 143–150 (2002)
2. B. Ahlgren, C. Dannewitz, C. Imbrenda, D. Kutscher, B. Ohlman, A survey of information-centric networking. IEEE Commun. Mag. **50**(7), 26–36 (2012)
3. A.M. Geoffrion, Generalized benders decomposition. J. Optim. Theory Appl. **10**(4), 237–260 (1972)
4. M. Hajimirsadeghi, N.B. Mandayam, A. Reznik, Joint caching and pricing strategies for popular content in information centric networks. IEEE J. Sel. Areas Commun. **35**(3), 654–667 (2017)
5. M. Mangili, F. Martignon, S. Paris, A. Capone, Bandwidth and cache leasing in wireless information-centric networks: A game-theoretic study. IEEE Trans. Veh. Technol. **66**(1), 679–695 (2017)
6. S. Puglisi, J. Parra-Arnau, J. Forné, D. Rebollo-Monedero, On content-based recommendation and user privacy in social-tagging systems. Comput. Stand. Interfaces **41**, 17–27 (2015)
7. K. Wang, H. Li, F.R. Yu, W. Wei, Virtual resource allocation in software-defined information-centric cellular networks with device-to-device communications and imperfect CSI. IEEE Trans. Veh. Technol. **65**(12), 10011–10021 (2016)
8. T. Wang, J. Blocki, N. Li, S. Jha, Locally differentially private protocols for frequency estimation, in *Proceedings of the 26th USENIX Security Symposium*, Vancouver, BC, Canada, August 2017
9. C. Zhao, Y. Guan, Data-driven risk-averse two-stage stochastic program with ζ-structure probability metrics. Available on Optimization Online (2015)

Clock Auction Inspired Privacy Preservation in Colocation Data Centers

6

Abstract

Data centers are key participants in emergency demand response (EDR), where the grid coordinates large electricity consumers for reducing their consumption during emergency situations to prevent major economic losses. While existing literature concentrates on owner-operated data centers (e.g., Google), this work studies EDR in multi-tenant colocation data centers (e.g., Equinix) where servers are owned and managed by individual tenants and which are better targets of EDR. Existing EDR mechanisms incentivize tenants' energy reduction. Such designs can either be gamed by strategic tenants or untrustworthy colocation operators for illegal gains. These serious privacy concerns stand as barrier preventing the tenants' participation in EDR. This chapter addresses such concerns by proposing a privacy-preserving and strategy-proof mechanism using the descending clock auction. Privacy is protected by implementing homomorphic encryption for aggregation through the clock auction, where operator can only know the aggregate of the tenants' values or bids but not their individual private values or confidential information submitted to meet the EDR. We evaluate the privacy and performance of this scheme by formulating descending clock auction, in which the amount of energy/price the tenants are willing to reduce for a given price/energy to meet EDR is protected.

6.1 System Model and Preliminaries

6.1.1 System Architecture

We focus on the emergency demand response scenario in the data colocation centers, where the tenants participate in the EDR and receive incentives in return for their participation in EDR. The system model is as shown in Fig. 6.1. The tenants continuously submit their bids in the descending clock auction to the colocation

© The Author(s), under exclusive license to Springer Nature Switzerland AG 2019   57
M. Pan et al., *Big Data Privacy Preservation for Cyber-Physical Systems*,
SpringerBriefs in Electrical and Computer Engineering,
https://doi.org/10.1007/978-3-030-13370-2_6

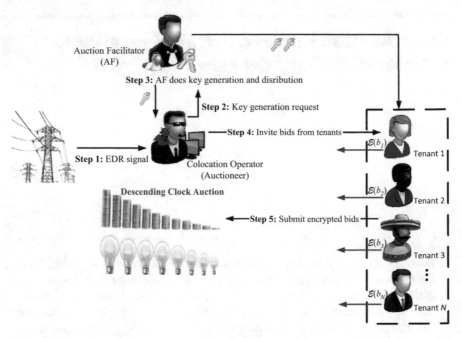

Fig. 6.1 System architecture of descending clock auction for EDR

operator until the target amount of energy reduction is reached and the total reward amount is under the specified budget. Suppose that the bidding values of tenants for EDR participation are $\mathcal{B} = \{b_1, \cdots, b_N\}$. For a price descending clock auction, we consider a data colocation center consisting of $\mathcal{N} = \{1, \cdots, N\}$ tenants, and a colocation operator; where the colocation operator invites tenants to send their bids for EDR participation on receiving an emergency demand response signal as shown in Fig. 6.2. The EDR signal specifies the target amount of energy that needs to be reduced in the colocation center. The colocation operator by means of a descending clock auction determines the winners when the total energy reduction target R_t is reached under a specified budget B_t. In each round k of price descending clock auction, the colocation operator announces the price offered per kWh of energy P_k and all the voluntarily participating tenants send the amount energy they are willing to reduce e_i for that price, and the value of e_i for tenants that aren't willing to participate in EDR is zero. In an energy descending clock auction, we consider multiple colocation operators trying to meet EDR instead of a single colocation operator. We consider an architecture with $\mathcal{M} = \{1, \cdots, M\}$ colocation operators and $\mathcal{N} = \{1, \cdots, N\}$ tenants distributed over these colocation data centers; where the colocation operators simultaneously invite tenants to send their bids for EDR participation as shown in Fig. 6.3. Similar to price descending clock auction, the colocation operators/auctioneers in the energy descending clock auction in each round k simultaneously announce the energy in kWh E_k that needs to be

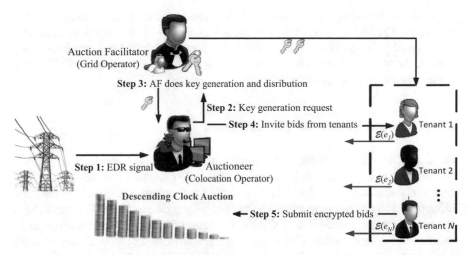

Fig. 6.2 System architecture of price descending clock auction

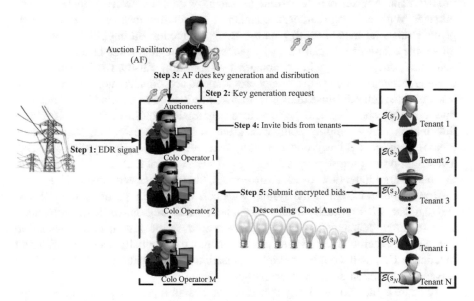

Fig. 6.3 System architecture of energy descending clock auction

reduced and all the voluntarily participating tenants send the reward price they are willing to receive s_i for reducing the announced amount of energy. As tenants can quote higher prices for energy reduction, to provide incentive compatibility in such architecture, the colocation operators provide a price p_j they are willing to pay to the participating tenants in EDR. The colocation operators decide on a clearing price c_{pk} in every round k based on the prices quoted by the tenants and all the operators, and the clearing price is paid to the tenants for energy reduction. Since the colocation operator is semi-honest in such auctions, we employ trustworthy entity as Auction Facilitator who jointly evaluates the winner with the operator to preserve the tenants' privacy. The auction facilitator is responsible for key generation and winner evaluation to have a secure descending auction for meeting the EDR while preserving the tenants' privacy.

6.1.2   Threat Model and Design Goals

During EDR, all the voluntarily participating tenants submit bids to receive incentives for their energy reduction. The tenants in the data colocation center may be some companies and the bids they submit may reveal the sensitive financial information about the companies. The tenants submit their private information b_i to the operator trying to meet the EDR, where in most of the cases colocation operators can be semi-honest, and disclose their private information to gain advantages. We assume an attacker wants to learn the private information of tenants from their bids in EDR. The attacker can be a strategic tenant or an untrustworthy operator or an outsider. Many existing market mechanisms (e.g., auction mechanism and pricing) provide financial incentives to tenants' energy reduction during EDR. These rely on simple solutions that can be easily gamed by strategic tenants or untrustworthy colocation operators to gain advantages by submitting falsified information. The existing mechanisms that are strategy proof are too complicated mechanisms, which ask tenants to fully disclose their private cost information (e.g., what is revenue loss due to energy shedding and performance degradation) during EDR. Such private information may be highly confidential, and soliciting it from individual tenants creates significant privacy concerns may preclude tenants' participation in EDR. Therefore, privacy-preserving strategy-proof mechanisms that can still incentivize tenants' energy reduction for EDR are needed, but these are very challenging.

The goal is to minimize the system-wide cost for efficient participation in EDR. Beyond that, EDR typically occurs throughout a wide region, while the colocation operator often manages multiple data centers over the region and tenants can flexibly route their workloads (and power demand, too) spatially over these different locations. This will couple multiple colocations' participation in EDR, making privacy preservation even more challenging.

To address the above privacy concerns, we design a privacy preserving scheme for meeting the emergency demand response. Since we assume the communication between tenants and colocation operator is not trustworthy, i.e., various adversaries such as eavesdroppers and tampers may be present. To have data privacy/tenant privacy, the data owner can resort to data encryption used in this scheme to encrypt the data before outsourcing to prevent the unauthorized entities from prying into the data. The bid integrity should be preserved, i.e., the bid information cannot be changed by malicious attackers or illegal competitors (strategic tenants). The untrustworthy colocation operator should not solely determine the winner since it may result in breach of tenants' privacy. Our major design goals to achieve a secure and privacy preserving EDR are as follows:

- The scheme should effectively achieve data privacy and bid integrity.
- The scheme should have a framework where an untrustworthy operator cannot learn tenants' private data or manipulate tenant's data.
- The scheme should achieve computation and communication efficiency compared with existing mechanisms.

6.2   Mechanism and Problem Formulation of PPCA for EDR

In this section we describe the problem formulation of PPCA for EDR. The descending clock auction [1–4, 7] used to meet the total energy reduction target is described in Sect. 2.3. We then give some background about the aggregation over encrypted data and its features. Then we elaborately describe our proposed privacy preserving data aggregation scheme for both price descending clock auction and energy descending clock auction cases.

6.2.1   Homomorphic Encryption for Aggregation

In this section, we describe the homomorphic encryption scheme used to aggregate the energy required to meet target energy reduction for EDR, along with the tenants' privacy protection. This scheme follows the idea of aggregation scheme designed in [8]. Our privacy preserving scheme uses this encryption scheme for aggregating the energy/price submitted by the tenants and check if the aggregated energy A_t meets the target energy reduction R_t.

Let $\mathscr{G}$ denote a cyclic group of prime order p for which the Decisional Diffie-Hellman Problem is hard to solve. Let $H : \mathscr{X} \rightarrow \mathscr{G}$ denote a hash function. A trusted authority chooses a random generator $g \in \mathscr{G}$ and random secrets $sk_1, sk_2, \cdots, sk_i, \cdots, sk_N \in \mathscr{Z}_p$. The public parameter is g. Each user i obtains a private key sk_i and the aggregator receives its private key $sk_0 = -(sk_1 + sk_2 + \cdots + sk_N)$.

- Encryption: During iteration k, user i encrypts its private value v_i as follows:

$$c_i \leftarrow g^{v_i} \cdot H(k)^{sk_i}.$$

- Decryption: Given the ciphertexts $c_1, \cdots, c_N$, the aggregator computes the aggregated sum of v_i as follows:

$$P \leftarrow H(k)^{sk_0} \cdot \prod_{i=1}^{N} c_i.$$

where,

$$P = H(k)^{sk_0} \cdot \prod_{i=1}^{N} c_i = H(k)^{\sum_{i=0}^{N} sk_i} \cdot g^{\sum_{i=1}^{N} v_i} = g^{\sum_{i=1}^{N} v_i}.$$

The aggregated sum of v_i can then be calculated by computing the discrete log of P base g.

6.2.2   Mechanism for Price Descending Clock Auction

In this section, we describe the formulation and overview of the first case of proposed scheme where the announced price decreases through the course of auction as shown in Fig. 6.2.

We illustrate the entire mechanism along with the problem formulation in three phases: Set up phase, Key generation phase, and Auction phase.

6.2.2.1 Set Up Phase
Set up phase initiates when the colocation operator receives the emergency demand response signal. The EDR signal specifies the target amount of energy to be reduced R_t under a specified budget B_t. The colocation operator on receiving EDR signal, simultaneously

- Invites bids from tenants to meet EDR, and
- Requests key generation from the Auction Facilitator who is trusted coordinator.

6.2.2.2 Key Generation Phase
In the Key generation phase, the Auction Facilitator generates the secret key for operator and tenants from random generator $g \in \mathcal{G}$ after it receives the Key generation request from the operator as follows:

- Tenants keys: Auction Facilitator generates random secret keys $sk_1, \cdots, sk_N \in \mathcal{Z}_p$ for every tenant i.
- Operator's key: Auction facilitator generates the colocation operator's key sk_0 as $sk_0 = -(sk_1 + sk_2 + \cdots + sk_N)$.

The auction facilitator then publishes the public parameter g, hash function $H(k)$ and distributes the keys to operator and tenants over a secure channel.

6.2.2.3 Auction Phase
The auction phase starts after all the tenants and operator receive their secret keys from the auction facilitator. Since we proposed DCA in the scheme, the bidder-specific prices start high and are decremented over the course of the auction. The auction phase constitutes of the following steps:

- Price announcement:
  In each round k, the auctioneer announces the price P_k willing to pay per kWh and invites for bids from tenants. The price P_k is highest for round 1 and decrements throughout the course of auction, i.e; $P_{k-1} > P_k > P_{k+1}$ for every round k.
- Tenants' bid encryption:
  In round k, each tenant i if interested in reducing their energy e_i for price P_k, sends its bid b_i encrypted with its secret key sk_i of amount of energy in terms of kWh to the operator.

$$b_i \leftarrow g^{e_i} \cdot H(k)^{sk_i}.$$

- Auctioneer energy aggregation:
  The auctioneer receives the encrypted bids from all the tenants and aggregates the total energy the tenants are willing to reduce in round k using its secret key sk_0 as

$$A_t \leftarrow H(k)^{sk_0} \cdot \prod_{i=1}^{N} b_i.$$

where

$$A_t = H(k)^{sk_0} \cdot \prod_{i=1}^{N} b_i = H(k)^{\sum_{i=0}^{N} sk_i} \cdot g^{\sum_{i=1}^{N} e_i},$$

$$A_t = g^{\sum_{i=1}^{N} e_i}.$$

The aggregated energy sum of e_i from the tenants can then be calculated by computing the discrete log of A_t base g. A_k is the aggregated energy in round k, i.e.,

$$A_k = \sum_{i=0}^{N} e_i = \log_g A_t.$$

- Auction continuation determination:
  The decision δ_k made by the auctioneer specifies whether the auction should be continued to the next round or not, based on the value of aggregated energy in round k, i.e., A_k and the budget for round k, i.e., B_k where

$$B_k = A_k \cdot P_k,$$

$$\delta_k = \begin{cases} k, & \text{if } A_k \geq R_t \text{ and } B_k \leq B_t, \\ k+1, & \text{if } A_k < R_t \text{ or } B_k > B_t. \end{cases} \tag{6.1}$$

The auction ends if the aggregated energy in that round meets the target energy reduction R_t of EDR and is under the specified budget B_t. The auction continues to the next round either if any of the above two conditions are not met. The auction then continues to the next round, the auctioneer decrements the price, invites bids from tenants for that price and determines the aggregate energy and the round budget. This repeats until the target energy reduction is achieved under the budget B_t.

- Winner determination and incentive determination:
  After the target energy reduction is met and is within the specified budget in a round k, the auction ends and the auctioneer sends the aggregated energy A_k of that round along with the individual tenant bids to the auction facilitator. The auction facilitator computes the aggregated energy similar to the auctioneer decryption from the tenants' bids and compares it with the aggregated energy value sent by the auctioneer to verify the auctioneer did not manipulate the data. The auction facilitator then decrypts the amount each tenant planned to reduce by computing

$$E_i \leftarrow H(k)^{-sk_i} \cdot b_i,$$

where,

$$E_i = H(k)^{-sk_i} \cdot b_i = H(k)^{-sk_i} \cdot H(k)^{sk_i} \cdot g^{e_i},$$

$$E_i = g^{e_i}.$$

Energy each tenant i planned to reduce e_i can then be calculated by computing the discrete log of E_i base g as

$$e_i = \log_g E_i.$$

The reward offered to each tenant P_i, that planned to reduce energy e_i for EDR is determined based on the winning round price P_k as follows:

$$P_i = e_i . P_k.$$

The auction ends and incentives are provided to all the tenants participating in EDR.

6.2.3   Mechanism for Energy Descending Clock Auction

In this section, we describe the formulation and overview of the second case of the proposed scheme where the announced energy decreases through the course of auction as shown in Fig. 6.3. This case completely differs from the price case since we consider a set of M operators/auctioneers planning to meet the EDR and a set of N tenants of these colocation data centers contributing to the EDR. To provide incentive compatibility in energy descending clock auction, the operators collaboratively determine the clearing price in energy descending architecture to provide incentives to tenants.

6.2.3.1 Set Up Phase

In set up phase each colocation operator invites bids from tenants to meet EDR after receiving the EDR signal, specifying the target amount of energy it needs to reduce R_{t_j} under a specified budget B_{t_j} receive the emergency demand response signal.

6.2.3.2 Key Generation Phase

In the Key generation phase, the Auction Facilitator generates the secret keys for operators and tenants from random generator $g \in \mathscr{G}$ after it receives the Key generation request from the operators as follows:

- Operators keys: Auction Facilitator generates random secret keys $sk_1, sk_2, \cdots, sk_j, \cdots, sk_M \in \mathscr{Z}_p$ for every operator j.
- Tenants keys: Auction Facilitator generates random secret keys $sk_{M+1}, sk_{M+2}, \cdots, sk_{M+i}, \cdots, sk_{M+N} \in \mathscr{Z}_p$ for every tenant i.

Auction facilitator generates the operators and tenants keys such that

$$sk_1 + sk_2 + \cdots + sk_M + sk_{M+1} + sk_{M+2} + \cdots + sk_{M+N} = 0$$

The auction facilitator then publishes the public parameter g, two different hash functions $H_1(k)$, $H_2(k)$ and distributes the keys to operators and tenants over a secure channel.

6.2.3.3 Auction Phase

In this case, the auctioneers/operators simultaneously announce the amount of energy that needs to be reduced and the tenants submit the price they are willing to receive for reducing that specific amount of energy in auction phase as follows:

- Price announcement:
  In each round k, the auctioneers announce the amount of energy E_k in kWh that needs to be reduced and invite bids from tenants in terms of price they are willing to receive in return for their reduction. The amount of energy E_k is highest for round 1 and decrements throughout the course of auction i.e; $E_{k-1} > E_k > E_{k+1}$ for every round k.

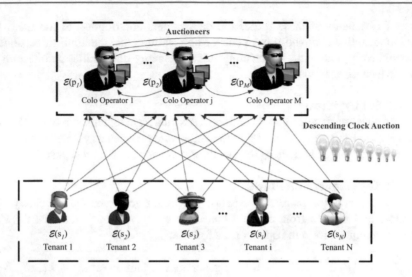

Fig. 6.4 Bid transfer in energy descending clock auction

- Operators' bid encryption:
  In round k, each operator j decides on a buying price p_j that it is willing to pay the tenants for reducing energy E_k and sends its bid b_j encrypted with its secret key sk_j to the remaining operators as shown in Fig. 6.4.

$$b_j \leftarrow [b_{j_p}, b_{j_e}]$$

where $b_{j_p} = g^{p_j} \cdot H_1(k)^{sk_j}$ and $b_{j_e} = H_2(k)^{sk_j}$,

$$b_j \leftarrow [g^{p_j} \cdot H_1(k)^{sk_j}, H_2(k)^{sk_j}].$$

- Tenants' bid encryption:
  In round k, each tenant i if interested in reducing its energy E_k for a selling price s_i, sends its bid b_i encrypted with its secret key sk_{M+i} to all the operators

$$b_i \leftarrow [b_{i_p}, b_{i_e}]$$

$$b_{i_p} = g^{s_i} \cdot H_1(k)^{sk_{M+i}} \text{ and } b_{i_e} = g^{e_{d_i}} \cdot H_2(k)^{sk_{M+i}},$$

$$b_i \leftarrow [g^{s_i} \cdot H_1(k)^{sk_{M+i}}, g^{e_{d_i}} \cdot H_2(k)^{sk_{M+i}}].$$

where e_{d_i} indicates the decision of either reducing the energy or not

$$e_{d_i} = \begin{cases} E_k, & \text{if willing to reduce energy in that round,} \\ 0, & \text{if not willing reduce energy in that round.} \end{cases} \tag{6.2}$$

- Auctioneer price aggregation:
  Every auctioneer receives the encrypted bids from all the tenants and operators and aggregates the total price the tenants are willing to receive P_t. The total price the operators are willing to pay P_o in round k using its secret key sk_i as follows

$$P_t = \sum_{i=1}^{N} s_i$$

$$P_o = \sum_{j=1}^{M} p_j$$

To aggregate $P_t + P_o$ operators compute
Operator 1: $\prod_{i=1}^{N} b_{i_p} \cdot (b_{1_p} \cdot \prod_{j=2}^{M} b_{j_p})$
Operator 2: $\prod_{i=1}^{N} b_{i_p} \cdot (b_{2_p} \cdot \prod_{j=1,3}^{M} b_{j_p})$ and so on
Every auctioneer computes the aggregate of $P_t + P_o$ as

$$P_t + P_o = \prod_{i=1}^{N} b_{i_p} \cdot \prod_{j=1}^{M} b_{j_p}$$

$$= g^{\sum_{i=1}^{N} s_i} \cdot g^{\sum_{j=1}^{M} p_j}$$

The aggregate of tenants' and operators' price in the round k can then be calculated by computing the discrete log of $P_t + P_o$ base g. $P_{tk} + P_{ok}$ is the aggregated price in round k i.e.,

$$P_{tk} + P_{ok} = \sum_{i=1}^{N} s_i + \sum_{j=1}^{M} p_j = \log_g P_t + P_o$$

- Auctioneer energy aggregation:
  After the auctioneer computes the cumulative price, it aggregates the total energy the tenants are willing to reduce in round k from the received encrypted bids as

$$A_t \leftarrow (\prod_{j=1}^{M} b_{j_e} \cdot \prod_{i=1}^{N} b_{i_e}) = g^{\sum_{i=1}^{N} e_{d_i}}.$$

All the tenants may or may not be willing to participate in EDR. If N_{tk} tenants are willing to reduce energy out of N tenants in that round k, then

$$\sum_{i=1}^{N} e_{d_i} = N_{tk} \cdot E_k + (N - N_{tk}) \cdot 0,$$

since e_{d_i} is zero if tenants are not willing to reduce the energy announced in that round E_k.

The aggregated energy sum of e_{d_i} from the tenants can then be calculated by computing the discrete log of A_t base g. A_k is the aggregated energy in round k, i.e.,

$$A_k = \sum_{i=0}^{N} e_{d_i} = \log_g A_t.$$

The tenants participating in EDR in round k is given as

$$N_{tk} = \frac{A_k}{E_k}$$

- Clearing price determination:
  Every operator computes the clearing price c_p to be paid to the tenants for energy reduction based on the aggregated price from operators and tenants. The operator determines the clearing price as

$$c_p = \frac{P_t + P_o}{N + M}$$

Since all the tenants may or may not be willing to participate in EDR, we consider the tenants willing to participate in EDR N_{tk} in round k while calculating the clearing price c_{pk} as

$$c_{pk} = \frac{P_{tk} + P_{ok}}{N_{tk} + M},$$

$$c_{pk} = \frac{\sum_{i=1}^{N} s_i + \sum_{j=1}^{M} p_j}{N_{tk} + M}.$$

- Auction continuation determination:
  The decision Δ_k determined by the auctioneers' decisions δ_{kj} specifies whether the auction should be continued to the next round or not. The auctioneers determine the δ_{kj} based on the value of aggregated energy in round k, i.e., A_{kj} and the budget for round k, i.e., B_{kj}.
  Based on the clearing price c_{pk}, each auctioneer determines the no. of tenants N_{jk} it requires to meet its EDR demand from

$$A_{kj} = N_{jk} \times E_k$$

$$B_{kj} = N_{jk} \times c_{pk}$$

and determine their decision as

$$\delta_{k_j} = \begin{cases} k, & \text{if } A_{k_j} \geq R_{t_j} \text{ and } B_{k_j} \leq B_{t_j}, \\ k+1, & \text{if } A_{k_j} < R_{t_j} \text{ or } B_{k_j} > B_{t_j}. \end{cases} \tag{6.3}$$

The decision to continue to the next round is determined from the auctioneers' decisions as

$$\Delta_k = \begin{cases} k, & \text{if } \delta_{k_j} = k, \forall j \text{ and } \sum_{j=1}^{M} N_{jk} \geq N_{tk}, \\ k+1, & \text{if } \delta_{k_j} \neq k, \forall j \text{ or } \sum_{j=1}^{M} N_{jk} < N_{tk}. \end{cases} \tag{6.4}$$

The auction ends if the aggregated energy in that round meets the target energy reduction of EDR and is under the specified budget for all the auctioneers. The auction continues to the next round either if any of the above two conditions are not met. The auction then continues to the next round, the auctioneers decrement the energy, invite bids from tenants for that energy and determine the clearing price, energy being reduced and the round budget. This repeats until the target energy reduction is achieved under the budget for all the auctioneers.

- Winner determination and incentive determination:

After the target energy reduction is met and is within the specified budget in a round k, the auction ends and the auction facilitator reveals the participating tenants in the winning round. The auction facilitator decrypts the price each tenant planned to reduce

$$P_i \leftarrow \Pi_1(k)^{-s_{kM+i}} \cdot b_{i_p}$$

$$s_i - \log_g P_i.$$

and reveals if the tenant participated in the winning round. Each operator sequentially chooses the tenants it needs to meet the EDR N_j from the participating tenants N_t. Then the reward of each tenant c_{pk} of winning round is provided as incentive to all the tenants participating in EDR.

6.2.4 Differential Privacy Preservation

We observe that in each round, the auctioneer needs to know only the aggregated sum of energy or price the tenants are willing to reduce or receive for participation in EDR. So this chapter implements aggregator-oblivious aggregation approach where the auctioneer can obtain the aggregated energy/price sum without knowing the actual energy/price value of each individual tenant. The aggregation process iterates until the target energy reduction is met under a budget. In the k-th iteration each tenant has private value b_i and the auctioneer needs the aggregate of b_i, i.e., $\sum_{i=0}^{N} b_i$ where b_i can either be a certain value of energy or price. The individual private data values of tenants are kept secret through encryption. To verify the identify of the tenant, the tenant also may sign the bid with its own private key and sends its public key to the auctioneer for its identity verification. For calculating the aggregated sum, if we use additive homomorphic encryption scheme like Paillier cryptosystem [6] for aggregation; where private data values of tenants are encrypted with the public key of auctioneer who can decrypt the aggregated ciphertexts. However, if the auctioneer gets individual ciphertexts of the tenants, it can directly decrypt and learn private data. Since, we assume the auctioneer, i.e., colocation operator, is untrustworthy we don't want the operator to learn the individual values. To prevent colocation operator from learning and manipulating the private data, we used the homomorphic encryption for aggregation scheme described in Sect. 6.2.1. This scheme preserves the individual privacy of the tenants while computing the aggregate of their private values, i.e., our scheme achieves the differential privacy preservation while meeting the emergency demand response. The auctioneer cannot trace back the actual private values from the bids it received guaranteeing differential privacy.

6.3   Security and Performance Analysis

In this section, we discuss the security properties and the performance analysis of the EDR data aggregation scheme.

6.3.1   Security Analysis

In this scheme, the confidentiality and data privacy are achieved by encrypting the tenants' bids or any other sensitive information using the homomorphic encryption scheme mentioned in Sect. 6.2. The bid integrity/tenant's privacy is protected from the semi-honest auctioneer/operator, since the operator using its key can decrypt the total aggregated energy of the tenants and cannot learn any information about the individual tenants values. The auctioneer cannot manipulate tenant's information since it doesn't have access to the tenants' secret keys. The operator sends the tenants' bids of the winning round to determine the charging price, the auctioneer cannot send false data to gain benefits because the auction facilitator determines the incentives to be paid to the tenants after decrypting the energy values they submitted. The operator cannot misinterpret data using its key, thereby preventing fraudulent operators from gaining advantages and preserving the tenants' privacy effectively. We compare the results with Paillier cryptosystem [6], a public key encryption scheme that is traditionally employed for the energy aggregation in EDR (PCS for EDR) as shown in Table 6.1. This scheme effectively preserves the privacy of the tenants and also provides forward secrecy through the modified homomorphic encryption.

Table 6.1 Comparison of
security level

Security requirements	PCS for EDR	PPCA for EDR
Confidentiality	✓	✓
Authenticity	✓	✓
Bid integrity from tenants	✓	✓
Bid integrity from operator	×	✓

6.3.2   Performance Analysis

In this section we analyze the performance of the proposed scheme based on the
computation and communication overhead.

6.3.2.1 Computation Overhead

In our proposed scheme for data aggregation preserving tenants' privacy, the
computation overhead is higher for the exponentiation operations in $\mathscr{G}$ with prime
order p than the hash operations. The computation overhead for hash operations is
almost negligible. In PPCA for EDR, for price descending clock auction described
in Sect. 6.2.2 each tenant requires two exponentiation operations in $\mathscr{G}$ and one
multiplication for encrypting their cipher texts. After the users send their data to the
operator, it gets the aggregate value which involves $N + 1$ multiplications and one
exponentiation, since we assumed having N tenants in the colocation data center.
The auction facilitator determines the incentives by N exponentiations and multiplications. We denote the computation overhead for multiplication operation and
exponentiation operation in $\mathscr{G}$ by C_m and C_e, respectively. For energy descending
clock auction described in Sect. 6.2.3 the computation is much higher than the price
case, since there are two bid groups pertaining to the operators and tenants. Each
bid has two parameters (b_{i_p}, b_{i_e}) and (b_{j_p}, b_{j_e}) that requires tenants and operators to
compute four exponentiation operations in $\mathscr{G}$ and two multiplications for encrypting
their cipher texts. After the tenants send their bids to the operators, they determine
the clearing price that needs to be rewarded from the aggregated price and the
energy being reduced by tenants, each of which involves $M + N$ multiplications
and two exponentiations. The auction facilitator determines the participating tenants
and provides the incentives to the tenants in winning round.

   In comparison, if we perform the data aggregation using PCS for EDR, the
operator should perform N decryptions of individual bids apart from the aggregate
for the price descending clock auction. The colocation operators need to perform
$M + N$ decryptions for energy descending clock auction apart from the aggregate.
Since, Paillier encryption is a public key encryption scheme, the computation time
for key generation and decryption is very much higher than our scheme which
generates keys from cyclic group of prime order p. Experiments were conducted
on a Intel Core (i7) 3.3 GHz system to study the execution time for both schemes
PPCA for EDR and PCS for EDR. The cyclic group is generated from prime p of
1024-bit and of order 160. The Pallier encryption and decryption keys are of length
1024-bits. The 1024-Paillier encryption and 1024-Paillier decryption cost 1.12 and

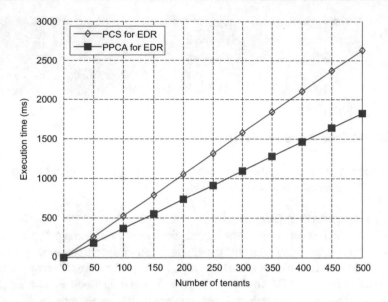

Fig. 6.5 Comparison of execution time in price descending clock auction

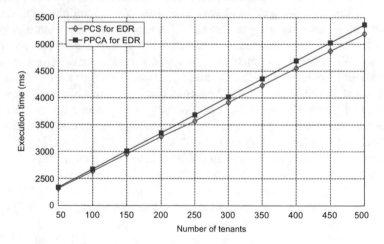

Fig. 6.6 Comparison of execution time in energy descending clock auction

3.58 ms, respectively. C_e in $\mathscr{G}$ for $|p| = 1024$ bits costs 0.28 ms and C_m costs 0.56 ms. The comparison of execution times is shown in Figs. 6.5 and 6.6, which compares the aggregation through PPCA for EDR and PCS for EDR schemes for both price descending clock auction and energy descending clock auction scenarios. In Fig. 6.5, we see that PPCA for EDR having two decryption operations at operator and auction facilitator achieves lower execution times compared to PCS for EDR for the price descending case. In Fig. 6.5, we see that PPCA for EDR has slightly higher execution times compared to PCS for EDR for the energy descending case despite having an additional bid parameters in two bid groups that increase the computation at tenants and operators.

6.3.2.2 Communication Overhead

In PPCA for EDR, the auction facilitator first generates the keys for all the tenants and colocation operator and distributes the keys to the tenants. The overhead for both PCS for EDR and PPCA for EDR is same in terms of key distribution, since the keys generated for both schemes are of 1024 bits. The overall communication overhead for PPCA for EDR relative to PCS for EDR is almost twice during the last round since the auctioneer sends the bids of the tenants to the auction facilitator for determining the incentives in price descending clock auction. The communication overhead is much higher in the energy descending clock auction due to the broadcast of tenants' and operators' bids to the operators. The communication overhead increases with the number of rounds in the auction as well as with increase in participating tenants and operators for EDR. To reduce the communication overhead, we optimize the price determination in the DCA [5] due to which the clock auction ends within a certain number of rounds and significantly reduces the communication overhead. We compromise the communication overhead for protecting the tenants' privacy from the operator which isn't achieved by the Paillier aggregation scheme.

References

1. O. Baranov, C. Aperjis, L.M. Ausubel, T. Morrill, Efficient procurement auctions with increasing returns. Am. Econ. J. Macroecon. **9**(3), 1–27 (2017)
2. D. Bergemann, B.A. Brooks, S. Morris, Optimal auction design in a common value model. Cowles foundation discussion paper no. 2064 (2016)
3. J. Levin, A. Skrzypacz, Are dynamic Vickrey auctions practical? Properties of the combinatorial clock auction. Technical report, National Bureau of Economic Research (2014)
4. D. Liu, A. Bagh, New privacy-preserving ascending auction for assignment problems (November 1, 2015). Available at SSRN: https://ssrn.com/abstract=2883867
5. T.-D. Nguyen, T. Sandholm, Optimizing prices in descending clock auctions, in *Proceedings of the Fifteenth ACM Conference on Economics and Computation*, Stanford, June 2014, pp. 93–110
6. P. Paillier, Public-key cryptosystems based on composite degree residuosity classes, in *Eurocrypt (Eurocrypt'99)*, Prague, Czech Republic, May 1999
7. R. Poudineh, T. Jamasb, Distributed generation, storage, demand response and energy efficiency as alternatives to grid capacity enhancement. Energy Policy **67**, 222–231 (2014)
8. E. Shi, H. Chan, E. Rieffel, R. Chow, D. Song, Privacy-preserving aggregation of time-series data, in *Annual Network & Distributed System Security Symposium (NDSS)*, San Diego, February 2011